奢尚新古典

欧朋文化 策划　　黄滢　马勇 主编

LUXURIOUS
NEO CLASSICAL

中国·武汉

CONTENTS
目录

传统新古典 / TRADITIONAL NEO CLASSICAL

006	现代巴洛克	006	Modern Baroque
018	究极新古典，艺术至高境界	018	Highest-Ranking Classic
028	精雕新古典，低调奢华风	028	Neo-Classical, Reserved and Luxurious
036	黑白灰的华彩演绎，极尽优雅尊贵	036	Distinguished Interpretation of Black White and Grey
048	新古典层峰，坐拥辉煌国度	048	Excellent Classic
054	精工美学，筑就传世府邸	054	A Mansion by Aesthetics
060	樱舞烂漫，梦幻古典御宅	060	Mythical Artistry
068	英式古典，细致优雅	068	British Classical Delicate
074	馥曼优雅，法式古典殿堂	074	French Classical Palace
084	安琪尔的梦幻城堡	084	Angels' Castle
096	欧洲经典，贵族格调	096	The Classic European, the Style Noble
102	瑰美殿堂，穿越浮生若梦	102	Magnificent Hall, Traversing Life Like a Dream
118	雍荣华府，世袭尊贵	118	Be Noble and Glorious and to Be
124	法式浪漫，精巧秀丽	124	French Romantic, Delicate and Beautiful
132	清新甜美，温情脉脉	132	Fresh, Sweet, Sentimental
138	精雕细琢，英伦华彩	138	British Resplendence by Great Care
150	怀想巴黎，女设计师的浪漫情结	150	The Complex of a Female Designer: A Paris Dream
158	浓情意大利，贵族新风尚	158	An Italian Flavor, A New Aristocratic Trend
168	简于形，奢于内，尊享艺术飨宴	168	Internal Simplicity, External Extravagance
176	舒雅怡人的生活沙龙	176	Salon for Daily Comfort

创意新古典 / CREATIVE NEO CLASSICAL

184	蝶儿翩翩舞过浪漫梦境	Butterflies Dance
192	低调高雅，黑色诱惑	Elegance Restrained Temptation Black
200	粗犷圆木，搭建爱的港湾	Love Harbor with Logs
210	原木，原乡，原生态	Crud Wood, Native Land and the Original Ecology
222	倾世蓝宝石，绝艳维多利亚	Wondrous Sapphire, Amazing Victoria
236	田园美境，法式乡村	The Pastoral of French Countryside
244	倾心自然，舒适回归	Passion for Nature, Comfort to Return
252	洗净铅华的优雅	Elegance to Be Natural
260	舒适典雅的贵气	Comfortable, Elegant and Luxurious
266	再现英伦精致品位生活	Reappearance of the Exquisite British Life
272	中西合璧，混搭创新	Innovation by the East and the West
276	英式下午茶	Afternoon Tea
284	和弦悠扬	Chords Melodious
290	华典悦章	Huadianyuezhang
296	一个空间，两种表情	One Space, Two Expressions
304	跨入新式奢华疆界	Luxurious Feeling

传统新古典

现代巴洛克
Modern Baroque

设计公司：玄武设计群
设计师：黄书恒
面积：100 m²
主要建材：酸蚀灰镜、黑云石、银狐石、土耳其黄、墨镜、银箔

Design Company: Sherwood Design Group
Designer: Huang Shuheng
Area: 100 m²
Materials: Etching Gray Mirror, Black Marble, Silver Fox Stone, Turkey Yellow, Sunglasses, Silver Foil

华丽繁复的线条之舞

勒·柯布西耶 (Le Corbusier) 在《迈向新建筑》一书中曾说："建筑是量体在阳光下精巧、正确、壮丽的一幕戏。"对玄武设计而言，室内设计也是一个由艺术元素、材料质感和视觉节奏所表达的一门剧场学，我们常像个安排空间的导演，让空间富于戏剧效果，在内隐-外显、收迭-张放、静止-行动间，塑造戏剧张力，营造令人惊奇的空间奇趣。

巴洛克风格的特征是华丽、力量、富足，服膺着17世纪的欧洲，向外扩张，追求财富的时代氛围。一方面发展科学，同时也因为不断征战而动荡，故巴洛克风格喜用繁复、富丽的流动线条表达强烈感情。玄武设计掌握其中艺术精神，去芜存菁地以黑、灰、白为色彩基调，加上少量金、银勾边与装饰，辅以亮面材质、水晶、玻璃产生的光影，用视觉动静的极度反差，激荡出新奇前卫的巴洛克美学。

嘉年华式的感官欢愉

赏析本案，如同观赏一出以浮华人生为主题的超现实歌舞剧，提供观者突破框架的想象力、混合梦境与现实的虚幻效果，以及强烈反差形成的戏剧张力，藉由线条、图腾、装饰与家具层层开展，传达空间的丰富动感，让每一位参访者随着空间铺陈而舞在其中。

空间要素如同嘉年华会的狂欢舞者，以造型装扮抢夺目光，舞出感官欢愉。客厅的银色雕柱与黄金纹饰，雪白圆柱与绸缎布面，简约与繁复于此并行不悖；玫瑰花形垂下的水晶吊灯，光影洒落于雕饰之间，营造出现代巴洛克的华丽和沉静；设计者利用酸蚀灰镜技术，使墙面浮出花草图饰，远观流泄一股静谧之气，近看却能令人惊喜再三；凹凸浮雕背墙、壁炉电视柜、门片与柱廊等处，以黑白两色石材，将浮华巧妙地转化为优雅。

空间细节充满巧思，如法式布帘与纱帘的倒置、精雕细琢的鞋柜把手、金色小孩的灯具、Ghost 的经典设计椅与圆柱雕饰的镜面倒影，让人处处惊喜，犹如嘉年华会中不时出场的诙谐角色，将气氛炒热到高点；黑白棋盘地坪是嘉年华会的大舞台，让所有角色轻盈跳跃，流连忘返，终至醉卧在这场巴洛克盛会中。

浮华人生的细腻沉思

这场"超现实"的巴洛克展演，是设计者对于现状的嘲讽。在房产的泡沫游戏之中，人们对于住宅形式的夸张演出浑然不察，设计者有意将空间设计作为舞台，施展对于虚假现实的基础抵抗；这出「雅俗共赏」的空间大剧，同时也是设计者在艺术性与现实的商业需求间，企图取得的最大平衡，即便是必须极度夸耀设计手法的售楼处，也要运用元素持续创造惊叹，让感受突破框架限制，正如"玄武"兼具蛇的灵动与龟的踏实，玄武设计未来也将秉持强大的创意以及踏实的执行能力，持续追求心中最崇高、最伟大的建筑空间。

Gorgeous and complicated line dance

Le Corbusier Le Corbusier, towards the new "building" a book once said: "Architecture is the amount of the body in the sun, compact, accurate, spectacular drama of the scene." Basaltic designs interior design is also an expression of artistic elements, materials, texture and visual rhythm of a theater, we often like the director of the arrangement space, space full of dramatic effect, implicit - the explicit closing Diego-Zhang Fang, still-action , shaping the dramatic tension, and create amazing space Trolltech.

Baroque style is characterized by beauty, power, wealth, adhering to the 17th-century Europe, to expand outwards, and the pursuit of wealth era atmosphere. because of the development of science, and the continuous campaign and turbulent, the Baroque style is like to use a complicated, rich and beautiful flow lines to express strong feelings, the basaltic design master which artistic spirit, the wheat from the chaff in black, gray, white color tone, plus a small amount of gold, silver crochet and decoration, supplemented by the bright surface material, crystal, glass, light and shadow, bring out the visual movement of the extreme contrast between the new and avant-garde Baroque aesthetic.

Carnival of sensory pleasure

Appreciation of the case, like watching a surreal musical, flashy life the viewer breakthrough in the framework of the imagination, mixing dreams and reality of illusory effects, as well as a strong contrast to the formation of dramatic tension by line, totem, decorative layers and furniture carried out, to convey the rich dynamic of space, space lay out and dance in which every visiting.

Spatial elements like carnival dancers snatch eyes to the shape of dress and dance of sensory pleasure. Silver carved columns and gold decoration of the living room, white cylindrical silk cloth, simple and complicated this go hand in hand; rosettes hanging crystal chandeliers, light and shadow falling on the carving between, creating a modern Baroque's gorgeous and quiet; designers take advantage of the etching grey mirror technology, wall floating flowers and figure ornaments, distance, stemming the flow of air of relaxed, look at the past was able to surprise repeatedly; bump relief back wall, fireplace TV cabinet door tablets and colonnade, etc. black and white stone, will be flashy and clever elegance.

Spatial detail is full of ingenuity, such as the inversion of the French curtains with gauze, crafted shoe handles, the lamps of the golden child, the Ghost of the classic design chair and carving cylindrical mirror reflection, people everywhere surprise, like a carnival played humorous roles from time to time, the atmosphere sizzling to a high point; the Othello plate floor is the big stage of the carnival, light jumping, forget all the roles, and finally to Zuiwo this Baroque event.

Delicate contemplation of the vanity of life

This "surreal" Baroque performances, is a mockery of the designers for the current situation, in residential form exaggerated performances totally be aware of the real estate bubble game itself, the designer intends to design the space as a stage to display for false realistic basis to resist; "tastes" a large space drama, but also the designer maximum balance between the artistic and practical business needs, attempting to obtain, even to extreme boast the sales offices of the design techniques, but also use of the elements continue to create amazing, feel a breakthrough framework constraints, both the snake's clever and turtles as basaltic, basaltic design of the future will also uphold the strong creative and practical ability to execute, continue to pursue the hearts of the most noble, the great architectural space.

In Search of Eternity
黄书恒建筑师·玄武设计隽品集

"黄书恒作品中'神圣'与'媚俗'的反复交辩，是他对人生最强劲的探问。"
　　　　——邱德光（新装饰主义大师）

"黄书恒展现了将机械科技导入空间设计的特殊美学。"
　　　　——阮庆岳（元智大学教授、知名建筑评论家）

邱德光、阮庆岳／联合推荐

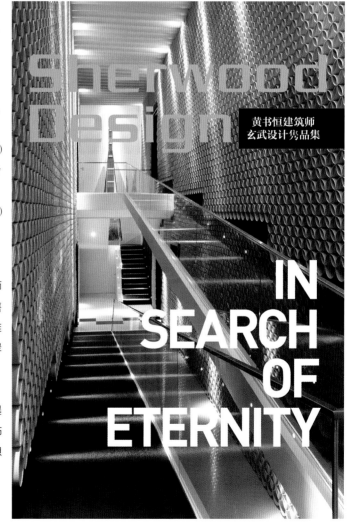

作为台湾最大建筑商——远雄集团的首席合作设计师，台湾知名建筑师、设计师黄书恒参与远雄集团于新北市的系列造镇计划所战皆捷，成功为业主创造漂亮的销售业绩，从户型平面布局到样板房、售楼处、公共会所，一条龙式的服务成功塑造远雄品牌形象。近年更将触角深入中国大陆，成为高端房产的设计主力，其作品成为两岸三地业主的营销利器，堪称营销策略最有力之推手。

《In Search of Eternity》即为黄书恒于2012年推出的新版作品集，涵括未曝光的精彩个案。包括荣获2011年日本JCD设计大奖之"金华苑售楼处"，现代装饰国际传媒奖-年度最佳展示空间，同时获观众票选为最佳展馆之"台北国际花博梦想馆"等重要案件，呈现黄书恒对于设计领域深层的反思与实践。

究极新古典，艺术至高境界
Highest-Ranking Classic

设计公司：古钺设计
设计师：颜国州
面积：400 m²
主要建材：天然石材、天然木皮、复古砖、订制家具、皮革、香槟银箔、鸵鸟皮、壁布、珍珠母贝

Design Company: Gu Yue Design
Designer: Yan Guozhou
Area: 400 m²
Materials: Marble, Wood, Retro Brick, Custom Furniture, Leather, Champagne Silver, Ostrich Skin, Wallpaper, Mother of Pearl

这是一个让心灵产生深层震撼的作品，设计师颜国州将新古典空间的每个角落、每个细节，都当成独一无二的艺术品来呈现的例子，虽然是再真实不过的生活空间，却又如此似梦似幻，让人过目难忘。

追求空间艺术的巅峰之美

宛如将值得永世典藏的古典宫廷挪移到了眼前，古钺设计将一贯坚持的极度细腻的艺术精雕精神，再一次发挥得淋漓尽致，并将每日进行的真实生活，转化为难以取代或模仿的空间艺术。

走进玄关，精雕细琢的工艺精华让人不由得发出赞叹，右侧三列立体收纳柜设计，造型上已经让人大感不可思议，设计师还在天然木皮上点缀栩栩如生的鸢尾花图案，堂堂宣示与众不同的豪门风采。地面圆融石材拼花，对应上方灿烂水晶灯饰，华贵的光影映照在气象辉煌的正面端景上，以多层次几何线条层叠并精工贴覆香槟银箔的不凡质感，彰显细节处的讲究与艺术层面的精益求精。

走过工序极度繁复、象征空间转换的造型拱门进入客厅，壮阔的殿堂气度令人忍不住要深呼吸，整个公共空间地面全数铺设烧面复古砖，并在客厅区域拼出地毯般的瑰丽图案。客厅天顶正中安排形状如太阳光环的立体天花造型，宛若现代抽象雕塑艺术的沙发背墙，以俯瞰盛开的玫瑰花瓣为灵感，繁复与精致绝伦的高度工艺，见证独一无二的艺术价值；下方以纯白大理石雕刻兼具实用价值的精美壁炉基座，与墙面艺术雕刻造型相映成趣。

客厅主墙的处理尤其费时旷日，两侧景色优美的对称开窗，让中央深色石材打造的立体边框层次更为鲜明，中央等比四扇方格滑门，分别使用双色珍珠母贝细腻镶嵌，门片后方分别是居中的电视墙与两侧精品柜，柜内包覆鳄纹皮革营造的低调奢华感，在灯光的烘托下尤其出色。

衣香鬓影的梦幻宫廷

走过美仑美奂却又隐约洋溢着休闲气息的气派客厅，左边是体态优美、蜿蜒向上的特制扶梯，右侧则是另一重无法言喻的感官洗礼。穿过两边宏伟的精雕罗马柱，来到两边规划落地大窗，可随时欣赏自家庭园美景的明亮宴客厅，厅内可供多人同时餐叙的巨大长桌，由经验丰富的巧手师傅量身订制，居中的天顶上方以精致的圆拱天花板，搭配雅致古典吊灯，酝酿缤纷浪漫的光影。对应餐桌的电视墙以天然石材勾勒直列的立体层次，两侧圆顶拱柱造型掌握最佳视觉比例，在空间中画出连续的优雅圆弧，深邃的殿堂之美，光是视觉的赏心悦目，就足够令人回味了。

为了避免在如此讲究的新古典大宅中仍有开放厨房直接面对公共厅区的突兀感，设计师颜国州特地在餐厅转角处安排一座相当别致的角圆型吧台，作为厨房工作区的前缘，同时也让厨房有机会共享外面的无边绿意。吧台造型结构富于立体感，交错的外框以香槟银箔精工修饰，内衬黑色块面搭配鸵鸟皮，强调与众不同的精品质感。

登峰造极的工艺水平

顺着梯口卷贝造型的优美梯线缓步向上，光润的台阶均以上选石材打造，顺着动线依序向上，仿佛踏在浪漫的梦境里。

登上二楼时，开展在眼前的又是由衷的赞叹。这里规划主卧室和小孩房，开阔的廊道分别向左右延展，设计师以细致框状边框、内嵌灯光、行列立体层次的天花造型，对应地面的双色木地板块面拼花，营造行间的视觉变化与律动感。面向梯口的大型精品展示柜，则是动线风景的高潮戏，精致无比的几何方格之间，所见皆是针对这个空间特制的线条、雕刻图腾，方格内衬的木皮材质，甚至还精心织上华贵的金银丝线，好对应整座柜体经纯手工多次涂装施作，才能营造出的仿象牙雕刻质感，细节之讲究足见主人非凡的品位与艺术涵养。

满是丰盈古典魅力的主卧室内，和谐的大地色彩加上饱满、丰富的线条，雕凿出三度空间无与伦比的艺术美感。包括气势十足的四柱大床与床尾脚凳、床畔的灰金色梳妆台等，全数都是纯手工打造的独家精品，特别是四柱与足部象牙雕刻的独特质感，细腻且带着耐人寻味的温度，让这休憩空间的美，更多了些醇厚的人文气息。

对优秀的设计师来说，能获得慧眼识英雄的屋主在设计概念和预算上的双重支持，是件多么难得又开心的事，当然更要全力以赴！眩目的繁华稍纵即逝，只有真正的艺术才能滋养心灵，无畏时间检验。

A project this space is with shocking power to overspread new classical details everywhere to render a unique work of art. True as it is, it's dreamy, magic, and memorable.

The pursuit of spatial aesthetic

The neo-classical palace seems to have been presented into eyes, when spirits of art essence are realized concrete into every space, whose artistic quality is hard to take place and copy.

The entrance strikes and imposes an impression of fine craftsmanship. The storing cabinets of three rows on the right the essence of the process, people could not help but give praise, and the right are modeled incredibly. The natural veneer embellished with lifelike iris pattern is a dignified declaration. The flooring stone mosaic, corresponds to the crystal lighting above. The front end view, once with light cast on, is brought out an extraordinary texture with multi-level geometric lines and champagne silver foil, highlighting the attention to details and the refinement of the artistic level.

The arch extremely complicated, a symbol of space conversion, leads into the living room, whose tolerance is breathtaking. The entire ground of the public space is coated in retro brick, and magnificent pattern appears in the living area, above which the three-dimensional ceiling is like corona. The backdrop of the sofa is by great technologies and in its abstract sense, the design of which is blooming rose petals. The fireplace

base is carved with marble, functional and aesthetic, responding to the carving decoration on the wall.

The processing of the main wall in the living room takes a couple of days. Symmetrical windows on both sides take good views, enhancing the layers of the central three-dimensional frames of dark marble. The four sliding doors of equal ratio in the same place are respectively decorated with two-color mother of pearl. Behind the door are the TV wall in the center and the boutique cabinets on both sides; inside the cabinet body is crocodile-patterned leather to create a low-key sense of luxury, particularly remarkable in light.

Court

In the rear of the parlour, a specially-made stairs wind beautiful upward on left, while the right allows for a sensory baptism beyond description. Lines of carved Roman columns, guide to French windows, where there is banquet hall, transparent to enjoy garden views. The large long table is customized, above which the circle ceiling is equipped with classical chandelier, brewing colorful and romantic lighting. The TV wall corresponding to the table with natural stones outlines three-dimensional levels. The arch and pillar ceiling at a precise ratio, draws continuous elegant arcs, from which, the depth of hall, is eye-comfortable and memorable.

A circle bar counter at the corner of the dining room grounds off the towering sense by placing the open kitchen opposite to the public hall. That's an outpost of the kitchen and exposes it to the greenery outside. The modeling is solid, while the interior of the staggered frame is lined with champagne silver foil, as well as black ostrich skin to stress a distinctive texture.

Technical level

The stairs is designed into a rolling shell, whose pedals are of stone, and steps along which seem to lead into a dreamy world.

The remarkable 2nd floor harbors the master bedroom and children. rooms The corridor there extend broad to right and left. The ceiling geometric and built in light echoes with wood flooring of double colors to make vision changes and rhythm. The large-scale display cabinet facing the mouth of the stairs is where the most beautiful view by lines: the check is exquisite; lines and totems are specially made; the veneer liner is woven gold and silver silk yarns to be parallel to the hand-painting process of the whole cabinet body, just for the texture creation of antique ivory carving. Unintentionally, details confide in the owner's taste and art cultivation.

The master bedroom is abundant in classical charm, where the earth color is added with lines to sketch a three-dimensional unparalleled artistic beauty. Including the footstool and golden dresser, all items are purely handmade. The four-poster beds, in particular, enhance the mellow atmosphere of culture with the unique texture of ivory carvings.

For a good designer, it's lucky and pleasing to design a space whose owners have discerning eyes, and then provide double support in terms of design concept and budget, because only with a heart that has been nourished, the prosperous that is likely to fleet can be everlasting.

精雕新古典，低调奢华风
Neo-Classical, Reserved and Luxurious

设计公司：风动设计
设计师：何三泰
面积：660 m²
主要建材：多种大理石、吊灯、进口壁布、进口绒布、金银箔、皮革、镜面不锈钢、钢琴烤漆、烤漆玻璃、镜片喷砂、进口五金、无毒低甲醛浮雕地板及环保素材

Design Company: Pneumatic Interior Design Co., Ltd.
Designer: He Santai
Area: 660 m²
Materials: Marble, Chandelier, Wallpaper, Velvet, Foil, Leather, Stainless Steel Mirror, Piano Baking Varhish, Painted Glass, Sandblasted Mirror, Hardware, Low Formaldehyde Embossed Flooring, Environmentally Friendly Material

以新古典主义风格为设计基石的本案，得益于设计师运用一种多元化的思考模式，将怀古的浪漫情怀与现代人对生活的需求相结合，打造一个兼具华贵典雅与时尚个性的豪宅大户，让业主为满屋的古典气息所环抱的同时亦能充分感受到来自传统的历史痕迹与文化魅力。

一进门的玄关区光影绚烂，过人的精致性令人难忘，重要的收纳集中在右侧镜墙后方独立的衣帽间内，地面精湛的棋盘格拼花，分别倒影茶镜立面与衔接的钢琴烤漆处理面上，细节的精致让人倍感礼遇。

室内面积广达660平方米，公共空间呈扇型展开的宽幅视野，将偏英式的新古典美学发挥到了极致。开放规划的客、餐厅地面统一铺设天然石材并采用无接缝处理，并以特调的咖啡、灰、金、银等色阶串连全宅的尊贵色彩布局。客、餐厅间开阔的视野，隔着手工精湛的几何订制腰柜分野，辉映餐厅可容纳十人以上的大型餐桌椅配置，尤其能烘托非凡的尊荣价值感。

全案虽然受限于建筑的原有高度，不过设计者顺应梁向，精心打造精美的行列式分区天花造型，将丰富的视觉层次与间接光和谐的分割比例合二为一，瞬间释放大宅应有的辉煌气度。私密空间一律采用全套房式的尊荣配置，主卧室展现沉稳的名家气质，洋溢古典美的壁纸图腾与周遭动人的珍珠光泽，凸显低调内敛的奢华本质；男孩房则以优雅的个性化色系，创造空间艺术的全新境界。

A neo-classical project this space is, where a thought mode of diversity combines the romantic nostalgia into the modern life to create a luxurious, elegant and stylish mansion to expose the owners to the traces of history and from the traditional culture charm.

The entrance strikes an imposing and stunning impression gorgeous, exquisite and memorable. Behind the mirror wall in the rear stands a separate cloakroom. The ground is of exquisite checkerboard parquet. Treatment of the reflection of tea mirror for the facade and the piano stoving varnish is detailed to allow for a courteous reception.

In the internal 660 square meters, the public space expands in a fan form to carry out English neoclassical aesthetics to an extreme. The living room and the dining room are open, where the flooring is of natural stone seamless joint. With the hues of coffee, gray, gold and silver, the whole layout is strung. The geometric handmade cabinet there is exquisite to set off the configuration of tables and chairs to accommodate more than ten dinners. That's really a shading added around to make out the extraordinary honor.

Despite the limited height, the beam direction is made the best use of to craft a ceiling of exquisite determinant partition where visual hierarchies, indirect light and harmonious ratio are integrated to make an instantaneous release of a mansion brilliance and tolerance. Private space is fully facilitated: the master bedroom shows a calm temperament exclusive to an old famous family where classical wallpaper with totem and pearl luster highlight the essence of low-key and restrained luxury; the room for the boy is coated in elegant, personalized color to create a new realm of space art.

黑白灰的华彩演绎，极尽优雅尊贵
Distinguished Interpretation of Black White and Grey

设计公司：东易日盛家居装饰集团
设计师：孟也
面积：1 000 m²
主要建材：大理石、护墙、软包、艺术涂料等

Design Company: DYRS Design Group
Designer: Meng Ye
Area: 1,000 m²
Materials: Marble, Apron, Upholstering, Art Paint

本套案例是一套地下一层、地上四层的豪华别墅，业主是一位从海外归来的财富新贵，品位高雅。因此，设计师在此案例中摈弃了古典欧式繁杂细致的造型，专注于运用欧式线条及构造特点展现大气豪华的视觉空间，这种年代感与现代感的融合真正体现了业主对卓越生活品质的追求。现代、欧式元素的整体揉合，黑、白、灰的简单过渡让家居在闪烁的光影中更见优雅。人性化的家居，奢华而绝不浪费，简约而绝不被潮流所抛弃。

穿过铺满大理石悬挂羚羊头的玄关，步入大厅，在一个如画布般纯粹干净的黑白空间里，挑高的6米中空气势恢弘，简洁大方的黑白大理石与整个墙面有机结合，方形的天花穹顶大、中、小三层祥云式金属内架造型，冷金属流苏从房顶垂直泻下，拉伸整个客厅的空间感，与墙壁上金属碎片拼接的艺术品相趣成章。整个地面全部运用了淡黄色纹理白色大理石饰面，冷金属与冷色调的碰撞，表达出简约的内在，使得整个空间富有低调气质又不失味道。装饰的使用上都有欧式遗风，它们不仅造型华美，还增添了温婉的气氛。淡蓝色的古典欧式沙发，深蓝色的落地幔帐，是这里不可或缺的精彩演出，融合又跳跃充满趣味。

客厅旁边是岛台分割的厨房及餐厅，餐厅依然是黑色冷金属的流苏灯饰与客厅相得益彰，缓缓衍生，3米长的原木长桌置于正中，既保留了欧式古典风格中的高贵优雅，又吸取了现代简约风格的明快清爽，去繁就简尽显灵动之美。造型落地台灯给整个餐厅增添了一抹靓丽的红色，强烈的对比尽显脱俗的气质，华美的鹿角椅更是在典雅气氛中注入了一丝野性。

地下花园让那些古典情结尚未解开的人们可以在这里找寻潇洒的旧日痕迹。朴素的色彩，显得深沉静谧。简洁流畅的线条，精妙丰富的布置，在尽享功能和舒适的同时，处处流露出主人优雅从容的气质。这里没有多余的色彩，没有喧嚣与繁冗，一派宁静悠远。各种植物排列有序，共同营造休闲放松的场所。轻轻品茗一口，味道尽在其中。尤其是夜晚独步来到这里，仰望星空，投影灯打在墙壁上的大雁，更是意境非凡，回味深长。又能从花鸟鱼虫的精雕细琢中，体会到一种沉稳淡雅、修身养性的生活态度。

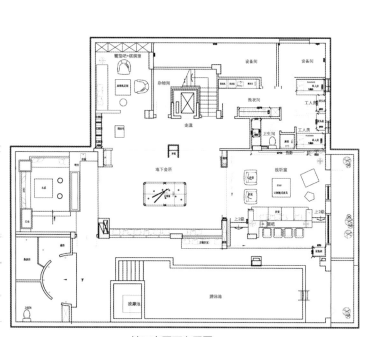

地下室平面布置图

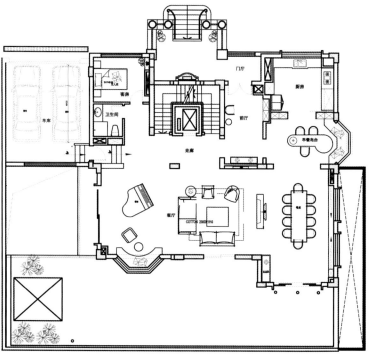

一楼平面布置图

二楼主卧极具层次感的软包皮面墙体，加以少许蓝色的点缀，充满了朝气，同时拥有了生机。形成了自然的色彩过渡和明暗区分，简约中透着华美和精致。偌大的旋转电视一体柜，使整个主卧加以适当区分，形成一个隔而不断，分而不离的互动空间。撇掉欧式家具的贵族气质，依然可以看见熠熠生辉的高调装饰，豪华但不浮夸，大气但不失精致！

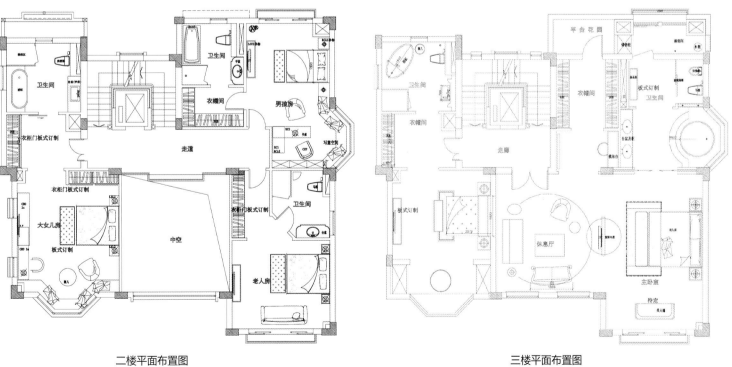

二楼平面布置图　　　　　　　　　三楼平面布置图

A luxurious villa this space is with a basement and four stories up-ground. The owner, a returnee and nouveaux, has tastes. Complex and detailed modeling of the classical European-style is therefore abandoned, but attention is paid to the feature shaping of lines and structure to make a visual luxury. The blend of the historical and the modern is actually reflects the owner's pursuit of excellent life. Elements modern and classical are fused. The transition black, white, and gray complements the elegance of the furniture with light and shadowing dancing. The home furnishing is humane, luxurious but not a waste. Just as testified, by no means is a simple design discarded.

The entrance is fixed with marble flooring and decorated with an antelope head. The hall measuring 6 meters high makes a canvas image in the purity of black and white space. The wall coated is integrated with black and white marble. The ceiling dome is patterned auspicious cloud of large, medium and small size, with metal frame lined. Metal tassels in cold on tone down the ceiling boast the spatial sense of the whole space, while echoing with art work of mosaic pieces. With the collision between the metal and the white marble flooring, the connotation of the interior is brought out, and the space is highlighted reserved yet with something worth appreciation. The decoration, is not only physically magnificent, but also adds a gentle atmosphere. The classical European sofa of light blue and the dark blue landing canopy play a key role, when integrated within and seem to have made surprises.

Flanked with the living room are respectively the kitchen and the dining room. The latter is decorated with lighting of black metal tassels, with a long table of wood in the middle to keep the usual classical and elegance but in a neat appearance, and floor lamp in red. Such a strong contrast injects the space with a refined temperament. The chair is antler-shaped and immits a wild air.

Garden is located downstairs for people to further a satisfaction of the classical pursuit. Plain colors are of staidness, lines of simplicity and configuration is of excellence. The perfection functional and comfort everywhere reveals the owner's grace and leisure. No color can be surplus, and the peace and quiet submerges the bustle and hustle. Plants in order make a field for entertainment and relaxation. Wandering at night makes a particular image with looses rejected on the wall. From the pattern of animals and plants like flowers, birds, and fish flows out a kind of staidness as well as an attitude to cultivate moral character.

The master bedroom on the 2nd floor, is embellished with upholstered wall patched somewhere blue to allow for vigor and vitality. The color transition and the division clear and dark are made into being. The simplicity reveals a sense gorgeous and refined. The huge rotating TV cabinet properly separates the master bedroom, but not to cut off. When free of the European furniture, the whole space is still shining with the high-profile decoration, luxurious but not exaggerated, magnificent and sophisticated.

新古典层峰，坐拥辉煌国度
Excellent Classic

设计公司：古钺设计
设计师：颜国州
面积：265 m²
主要建材：天然玉石、天然石材、金银箔、琥珀镜、贝壳板、皮革、艺术雕花线板、丝绒、海岛型木地板

Design Company: Gu Yue Design
Designer: Yan Guozhou
Area: 265 m²
Materials: Natural Jade, Natural Stone, Gold and Silver Foil, Amber Mirror, Shell Plate, Leather, Art Carved Strip, Velvet, Island Wood Flooring

发源自欧陆的新古典，原本就是一门博大精深的空间艺术。在古钺设计新近完成的住宅规划案例，设计师颜国州纯青的专业素养，再一次得到尽善尽美的发挥，除了精雕细琢的建筑语汇；天、地、壁多处立面造型浮雕镂刻的线条之美，还有各种精选材质共构所展现的精致奢华，让每一方寸的细节都是令人叹为观止的精湛工艺。

新古典的精髓不仅是风格语汇与空间比例的完美和谐，更在与视觉、触觉上的精致震撼。设计师以个人跨足多重设计领域的丰富经验，再度将住宅空间，升华到了艺术殿堂的境界，华贵的金、银箔闪耀温润光泽，充满力挺感与抽象美的现代雕塑，成玄关区独一无二的端景，加上各种精润的大理石材、拱顶天花板、造型拉门等等的浮雕线条，这些元素如同盛会上的星光烁烁，汇集新古典艺术的精华，并将发誓精雕与现代的简洁线条、材质特色等相互融合，创造无一不是目光焦点，无处不令人赞叹的当代极品。

立体雕塑 再造玄关惊叹号

玄关一进门，强调边缘层次的天花板造型，与地面工艺复杂的石材拼花相互辉映，为了避免视线直入餐厅，迎面一座供餐厅、玄关两面欣赏的精美端景屏风，毫无意外地吸引众人目光，以金箔衬底的半直纹、半乱纹背景。交代低调奢华的尊荣内损，最上层如抽象浪头般的力挺雕塑，则将现代艺术忠实地引入生活。

气宇轩昂的迎宾客厅，全室内外闪耀着迷人光彩，多层次供顶天花板造型，因为周边间接光的烘托而更立体，居中的浮雕饰搭配典雅的古典灯饰，让实现感受最美的仰角视野。质感华丽的玫瑰金箔，在设计师的手里成了随心所欲的颜料，精心打造出六等分连续半圆拱状的沙发背景墙，整体施作上不仅需要精湛的工艺支持，也凹陷出独一无二的艺术性。接着电视主墙也是重点之一，基

底大面积铺陈的天然玉石，蜜蜡般的温润感与金箔素材相互辉映，奢华指数已然登峰造极，而电视墙中央宛如象牙质地；雕工精美的壁炉式电视柜，中央对开的门片以锻造铁件修饰，利用柔美的云纹线条刻划画龙点睛的聚焦效果，也充分展现高人一等的大宅气度。

处处惊艳的壮丽殿堂

极致奢华、工艺细腻的餐厅与玄关比邻，随处都有令人惊艳的风景，其中顺应现场樑向打造的立体天花板造型，结合灯光与优美的弧线，巧妙修饰了横过餐厅中央的樑，并与大理石地面的神色滚边线条相呼应，款式优雅的圆形古典餐桌椅，对话上方圆形银箔天花板，顺势强化了餐厅的交谊轴心，并在周边高潮不断的立面造型间，创造各有所归的技能属性。

餐厅周边轮番上演着每仑美奂的新古典风华，首先是餐桌旁以天然玉石对花打造的里面设计，当后衬的光源亮起时，鲜明的石纹就像一幅有生命力的天然画作般，瞬间璀璨夺目，造型墙对面是一座精工嵌如墙面的精品柜，刻意露出双足的技巧，表现亮体与墙面勾缝线条天衣无缝的结构之美，体现出主人的不凡品味。

主人最爱的书房内洋溢沉稳的贵族气息，书桌后靠线条丰富的造型墙，予人极大的想象空间，宛如现代雕塑的对开拉门气势宏伟，彰显设计者对于细节的讲究。

私密空间是优雅新古典的最佳代言，其中主卧室拱顶的天花板造型，由中央的雕花饰板向四方星芒状辐射，彰显向上伸展的高度感，床头墙面丝绒绷制的柔软强调和谐比例，与元宝形的床头饰板互为表里，床尾安排大型衣柜与专属更衣间，其中衣柜以珍贵珠母贝板打造，融合低调奢华与现代的线条分割，赋予空间无懈可击的精致品味。

在令人赞叹不已的现代工艺背后，全案的精彩，仰赖设计者对新古典风格不断精益求精、求新求变的态度，并能将多年累积的美学精华，熔铸于空间当中，为使用者真正创造寓艺术于生活的超完美境界！

Originating in European neo-classical, originally is a profound space art. In Gu Yue design newly completed residential planning case, stylist Yan Guozhou state pure green professional accomplishment, once again to get perfect play, in addition to sculpture, architecture vocabulary, the heavens and the earth, the wall facade relief enchase the beauty of the lines, there are various kinds of selected material co-constructed shows delicate costly, make every details of the heart is breathtaking craft.

The neo-classical essence is not only about a perfect harmony of style, vocabulary and spatial scale, but also about a visual, and tactile shock. This is a space where the expert designer has sublimated residential project into a realm of art gallery, where gold, silver foil and sculpture has made a unique end view in the entrance, which with the employment of marbles, arch ceiling, sliding doors, French carving and modern lines, overspreads everywhere spotlights of stunning contemporary works.

Solid sculpture

In the entrance, the ceiling makes a sharp and intended contrast with the flooring of complicated stone mosaic. A screen that can serve as end view for the dining room and the entrance, is bound to focus eyesight, whose backdrop of vertical and curly grains as well as the solid sculpture of abstract wave interprets a low-key luxury while introducing modern art into life.

The welcoming room is imposing, where the multi-level vaulted ceiling looks more solid against the surrounding light, the central embossed plaque and the classical lighting allows for a best perspective. The gorgeous gold foil is crafted into six equal consecutive semi-circulars to constitute the sofa backdrop. That's a demonstration of

a superb technology highlighting the unique artistry. The base of the TV wall is of natural jade in large are, which contrasts with the gold foil. The texture of the central wall is like that of ivory; the TV cabinet is designed like fireplace, whose doors are symmetrical and are forged iron pieces, and the cloud pattern of which makes an effect of finishing touch, fully indicating the superior bearing of the space.

Hall stunning

The dining room, adjacent to the entrance, allows for stunning scenery everywhere; the three-dimensional ceiling goes along the direction of the beam, making a clever modification of the beam and echoing with the dark trims of the marble flooring; the round classic dining tables and chairs continue a dialogue with the silver-foiled ceiling, strengthening the linkage with the dining room and creating functional properties with the solid surroundings.

In the dining room, around tables is facade design registered by natural jade. Once the light source comes to life, the distinctive stone vein is like a picture made by nature. Opposite to the wall, stands a cabinet embedded in the wall. That's an expression of seamless stitch of the volume and wall to set off the owner's taste.

Filled with calm nobility owner favorite study desk after by the line shape wall, to the great imagination, like a modern sculpture outside the sliding door imposing to highlight designers pay attention to detail.

The private space is a representative of the neo-classical. In the master bedroom, the vaulted ceiling seems to be going upward with the aid of the central carved plaque that shines like stars. The soft velvet around the bed head emphasizes a harmonious proportion. The bed head is ingot-shaped, while the tailstock is fixed with large wardrobe and exclusive locker room. The wardrobe is made of pearl oyster plate, which integrates low-key luxury and modern lines, giving an exquisite and impeccable taste of space.

Only in the pursuit of keeping improving, can the modern technology make an incomparable design. Such is this project. And a perfect realm it is to blend art into life.

精工美学，筑就传世府邸
Mansion by Aesthetics

设计公司：风动设计
设计师：何三泰
主要建材：大理石、吊灯、进口壁布、进口绒布、金银箔、皮革、镜面不锈钢、钢琴烤漆、烤漆玻璃、镜片喷砂、进口五金、无毒低甲醛浮雕地板及环保素材

Design Company: Pneumatic Interior Design
Designer: He Santai
Materials: Marble, Chandelier, Wallpaper, Velvet, Foil, Leather, Stainless Steel Mirror, Sandblasted Mirror, Hardware, Low Formaldehyde Embossed Flooring, Environmentally Friendly Material

谁说豪宅大户的筑造必须是金光熠熠、万顷琉璃才能彰显其尊贵殊荣的？本案设计偏偏不走传统路线，偏执地只以灰、银两大色系配合上乘质感的后期软装便神奇地把一个飞阁流丹的大器度居住空间展现于人前，使人眼前一亮。

自成一格的独立玄关区采用两进式设计，精美的对开格子门适度交代内外的分野，地面使用多种天然石材菱格拼花，打造非凡的精致意象。过道右侧安排美仑美奂的衣帽间，间接光带辉映层架滚边琥珀镜的奢华肌理，在灯光下低调的璀璨光泽尤其赏心悦目。

客厅整体色彩的搭配堪称一绝，包括铁灰、浅褐、银、粉灰等等，让画面呈现丰富的层次感与立体感，从这样的构图技巧当中便可深刻感受到设计师对于色彩与时尚的敏锐嗅觉。此外，客厅内精心摆设纯手工订制的新古典家饰，对应四方立面圆融饱满的多层次装置艺术。为了避免实体隔间阻隔视线，设计师巧妙利用家具来分野客、餐厅，恢弘的电视主墙精选温润的石材精工雕琢，将与生俱来的美丽纹理，融入利落的几何分割层次之中，创造出独一无二的客制化艺术。工笔细腻绝伦的沙发背墙造型尤其令人赞叹，精算比例的墙体中央以来自意大利的立体花砖搭配石材镶边，对称的两边以弧形的银箔立面，温柔环抱浪漫的镜面喷砂线条并内衬灯光，让不可思议的瑰丽影像魅惑所有挑剔的眼睛，积极贯彻新古典大宅中不可或缺的对称布局与高度的美学造诣。

餐厅背墙精选图纹富丽的古典壁纸作为视觉主题，融合"对称"与"比例框"的运用，展现气势与精雕细琢的视觉之美，同时也是十人大餐桌的绝佳背景。

主卧室床头以巨大弧面银箔外框建构主体，框格内外分别使用尊贵的缎面和珍珠光壁纸，以精工笔触彰显现代美学，并营造高贵舒适的睡眠氛围。

A mansion doesn't necessarilary have to be made with gold or glass. This space is a project that subverts the mansion stereotype with the aid of gray, and silver as well as decoration and upholstering to make a dwelling space come magnificent and shining.

The foyer is shifted into two sections, where folio lattice doors draw a proper line between, the flooring is of a variety of natural stones to create extraordinary fine an image, the cloakroom stands on right of the aisle, and indirect light contrasts with the amber glass that frames the shelf trim luxurious texture, the low-key luster especially being pleasing.

A demonstration the color match in the living room is, of iron gray, light brown, silver, pink, and gray to present a rich sense of depth and three-dimensional composition, and one to indicate the designer's keen sense of fashion and colors. In addition, the neo-classical furnishings are handmade and custom, corresponding to the multi-level art installations on four walls. Remarkably, the furniture also serves as distinction between the living and the dining rooms to avoid vision blocks by items. The TV wall is of warm stone that fuses the inherent texture into the simpler geometric segmentation, creating a unique custom art. The sofa backdrop is meticulous, and exquisite. The central is trimmed with Italian marble. The sides symmetrical and decorated with the curved silver foil embrace sandblasted mirror with lights built in to complete incredible images challenging to all discerning, while actively implementing the symmetrical layout and the aesthetic attainments, attributes indispensable to neo-classical mansions.

The backdrop in the living room takes classic wallpaper as a visual motif, which integrates the use of the symmetry and proportion to reveal momentum and crafted visual beauty while making an excellent background of the large dining table.

As for the bedside of the master bedroom, a huge arc of silver foil outlines, within which satin and pearl light wallpaper are employed to with fine craftsmanship highlight modern aesthetics, and create a comfortable sleep atmosphere.

樱舞烂漫，梦幻古典御宅
Mythical Artistry

设计公司：古钺设计
设计师：颜国州
面积：200 m²
主要建材：天然石材、金银箔、琉璃、皮革、艺术雕花线板、彩绘、海岛型木地板

Design Company: Gu Yue Design
Designer: Yan Guozhou
Area: 200 m²
Materials: Marble, Foil, Glass, Leather, Art Carved Strip, Color Decoration, Wood Flooring

许多人欣赏过古钺设计的作品，无不对其中的创作灵感与精华主题啧啧称奇，一样是新古典家居，为什么古钺设计总是能呈现如此与众不同的精致感和艺术涵养？而且随着不同作品的发表，每一次都能带来前所未有的感官震撼与无尽赞叹！设计师颜国州在访谈中表示："能让空间总是有独一无二的风情与精湛的工艺，仰赖过去生涯中积累的跨界美学经验，从时装、绘画、建筑到空间设计，我对艺术的深度涉猎，决定了我规划空间的深度与广度，也让屋主能真正拥有值得珍藏的新古典艺术居家。"

奢华绝景 绚烂中磊拓大器

一进门的玄关区光影绚烂，过人的精致选型令人震惊，华贵的金、银箔闪耀温润亮泽、充满立体感的对称圆铁雕花造型与纹彩流动美不胜收的玻璃屏风，成了玄关区光彩夺目的端景，加上地面品润的大理石材拼花图腾，描边线条优美的天花层次、立面造型的鎏金彩绘等等，这些元素如同黑天鹅绒上的灿烂星光，汇聚当代新古典艺术的精华，并巧妙融入部分东方、禅风、法式精雕等鲜明的风格语汇，在多元文明共鸣的波涛中，树立每一方寸都是细节的标杆，而这些令人叹为观止的精湛工艺，是一页不折不扣的当代传奇。

走出玄关即为餐厅，长轴延展的开放公共空间，将极致奢华的新古典美学，带到了超乎想像的至高境界。室内地面充溢铺设天然石材并采用无接缝处理，特调的咖啡、褐灰、金、银等色阶串联全宅，尊贵而细腻色彩布局，让我们看见设计者对于空间用色的敏锐。

客厅以石雕壁炉加上精工彩绘，打造融入两侧特制精品柜的大器主墙面，雕花铁件的细腻纹理，与形体堆成比例、外观手工彩绘的精致美相互辉映，深刻诠释尊贵奢华的艺术神髓。沙发背墙的处理尤其经典，引用了日式贵族御用

屏风的装置概念，四折屏风上以华丽金箔为基底，烘托特地邀请专业画家费时旷日所完成：樱花满开时的壮丽剪影，透过精心描绘落英缤纷的樱舞绝景，东方花鸟的写意、伴随着和风的高压调味，在法式精雕的皇室贵气中，无可匹敌的空间艺术真是让人悸动莫名。

宜古宜今 当代空间之最

欣赏过公共领域前所未有的辉煌气度后，沿着中央走道进入房间区，走道前端是以夹纱玻璃拉门界定的精致书房，拉门材质轻盈的透光性，巧妙将部分廊道纳入书房使用，无形中也让空间感更开阔。光影缤纷的走廊另有一番讲究，底部金色漩涡纹立体端景，呼应方格延展的石材拼花地板，将丰富的视觉层次与间接光和谐比例合二为一，瞬间释放大宅的不凡眼界。

体现皇室尊龙的卧室群，每一间都是精彩可期，其中主卧室贯彻新古典大宅中不可或缺的对称布局，在床头墙面融入丝质拉扣绷布与对称线条造型，在床尾左侧休闲视听区沿用古典壁炉语汇，左侧则是通往浴室的隐藏式门片设计，最特别的是视听区左侧长列皮革衣柜创意，弧形内缩的立体门片结合了皮革的质感与手感缝制工艺，两两分割的门片上还错落点缀着清冽梅蕊，工艺之精巧难得一见。两间孩房也展现细腻浪漫的名家气质，洋溢古典美的床头绷布造型虽然色彩有别，但动人的丝光质地与精致感相辉映，凸显无可取代的奢华本质；也从此颠覆空间艺术的至高境界。

Any project by Mr. Yan, would stimulate admiration for its inspiration. People frequently wonder why projects by him can be more likely to allow for unprecedented sensory shock and win praise Designer. Just as he is quoted, only experience accumulated, knowledge and reference to a span of fields covering fashion, painting, architecture, space design, and art covered can determine the depth and breadth of spatial planning, thereby presenting homeowners a neo-classical collection in a dwelling space.

Luxurious and magnificent

The foyer is gorgeous, delicate and stunning. Foil of gold and silver shines and glisters. The three-dimensional iron-wares of symmetry and the glass screen make a dazzling end view. The flooring is marble parquet; the ceiling is in layers with graceful lines; and the façade is gliding. All are starry against a velvet setting. Meanwhile, art of the neo-classical is cleverly integrated with vocabulary of the east, the Zen, and the French to make a harmonious state, establishing a benchmark everywhere. That's an amazing superb process, and a fully-fledged contemporary legend.

The dining room is adjacent to the entrance. The open public space extends further while exerting the neo-classical to a highest level beyond imagination. Natural marbles are seamlessly joined on the round Particular hues of coffee, gray brown, gold, and silver string the whole space. Such a distinguished and delicate palette is a glimpse into the keen use of all colors.

In the living room, the stone fireplace is carefully color-painted; the main wall is built in with boutique cabinets on both sides; the delicate texture of the carved pieces of iron makes a good match with their symmetrical and hand-painted body to interpret luxury and artistic flavor. The backdrop of the sofa refers to screens used by Japanese nobility. The four-fold screen is endowed with a base coated in gold foil. On the surface is magnificent

silhouette of cherry blossoms. In a French royal setting, images of eastern flowers and birds as well as elegant sense complete an unrivaled throbbing sense.

Ancient and modern

The central aisle is bound to lead to the rooms. In the front, there is a study defined by yarn-clipped glass sliding door, which also expand the study into the aisle, potentially allowing for an opener sense. At the other end, is a solid end view of gold swirl pattern to correspond with the flooring of stone parquet, while enriching vision and expanding indirect light, instantaneously releasing the extraordinary vision of the mansion.

All bedrooms reflects the royal statue, each a brilliant show. The master bedroom is symmetrical, where the wall around the bed head is decorated with silk stretch fabric and symmetric lines, and the left of the tailstock hides the door to the bathroom, while the right for video and audio entertainment involves language of classical fireplace. Beside the video-audio wall stands a leather wardrobe, whose doors are three-dimensional, curved, recessed and integrated with leather texture and treated with hand-sewing techniques. The doors, remarkably, are decorated with plum stamen. The two rooms for the children confide in a temperament by design masters, where even though colors for bed head stretch fabric are different, the mercerized texture and the exquisite sense highlight an irreplaceable luxury nature, and then the highest level of space art is subverted.

英式古典，细致优雅
British Classical Delicate

设计公司：奥迪国际室内设计
设计师：杜康生
面积：825 m²
主要建材：黑金锋大理石、深金锋大理石、灰姑娘大理石、旧米黄拼贴

Design Company: Audi International Interior Design
Designer: Du Kangsheng
Area: 825 m²
Materials: Marble, Collage

本案是华人富豪圈中最爱指定的设计团队—奥迪设计的又一力作，豪宅设计大师杜康生擅长将大尺度的住宅空间，设计出既贵气又气质非凡的尊贵豪邸。

本案从优雅的线面中，撷取轻柔淡雅的英式古典元素，勾勒细腻的空间姿态，奥迪设计将本案豪宅的器度，通过现代观点诠释，将古典的华丽与尊爵，于此空间内驰骋，随着线与面的开展、收合，擘划英式古典的设计精髓。

面对每一个在金字塔顶端的业主，奥迪设计杜康生设计师以设计者的敏锐与专业，解析不同个案，依其空间格局、动线、品位需求、生活习惯等等，汇聚丰富的空间概念，透过与众不同的设计，发展独特的空间场域。擅长以古典语汇成就毫宅量体的杜康生，因应本案业主需求，诠释细腻浪漫的英式古典。

甫入内随即以大理石墙与尊贵的顶级家具牵引空间气势，天花板以奢华的水晶灯饰呼应地面大理石材，双入口的玄关设计成就豪奢风范。业主对于材料的质感有严格的要求，因而挑选纹理极佳的地面做为空间基底，与设计师在设计上的专业相辅相成。

客厅以开阔的气势与玄关相互映衬，色调的运用，以精致典雅的米黄大理石铺陈整室，内敛的沉稳色系衬托空间里点、线、面的细致比例，随着细致的线条转折，展绎英式优雅的经典语汇。整体家具细致的材质透过不同灯距、光源，带出其经典绝美的的肌理，比例、颜色、软硬材质交错融合，织就英式古典庄严又优雅的华丽意象。以温润古典的 BAKER 餐椅，搭配自天花板垂坠而下的水晶吊灯，为餐叙空间挹注丽致质感。空间的规划上，厨房与餐厅之间利用简单的吧台设计界定空间，同质设色让整体视感呈现一致性，视觉宽敞通透。

延续公共区域的古典气息，主卧以淡雅的白色系呼应英式古典，并藉由材质的转化区分公共、私密空间，紫檀木地板的温润质感与白色的雕花壁面并蓄着眠卧空间的舒适甜美。主卧主墙透过典雅的裱布、对称的床头灯饰，以壁灯映照出空间层次，在静谧的古典氛围里展演优雅气韵。偌大的主卧空间，以镂空屏风区隔出睡眠区与起居室，减缓空间的局促，创造出隐而通透的空间层次。主卧卫浴足以展现极致奢华，全室以高贵的大理石雕刻而成，费时费工的石材折绕线条，与加大的洗手台面以及完整的收纳机能，让整体空间散发出英式尊爵的古典品位。

Another project by Audi Design, the most believable and favored team in the circle of Chinese tycoons, this space is, where Mr. Du Kangsheng, a master good at large-scaled mansion has successfully brought out dignity and grandeur.

English classical elements were captured gentle and elegant from elegancy of lines and surfaces, making a good sketch in a delicate space posture. Beneath the hands of Audi Design, characteristics of a mansion has been interpreted to the utmost in a modern approach: the classical, the dignified and the gorgeous, as lines extend and surfaces expand, have been running throughout to set off the essence of the British classical.

As usual, for individuals above the top of the wealth pyramid, Mr. Du would be professionally and keenly devoted to presenting projects that are well

consistent with spatial pattern, lines, needs, tastes and living habits. And this project, in line with the personal requirements, is a delicate and romantic interpretation of the English classical.

Directly from the double entries that contribute to an extravagant style, space momentum is ignited by marble wall and top furniture, where ceiling with its luxury crystal lighting echoes with the marble of the flooring. The careful selection of flooring marble, not only meets the owner's strict requirements for the texture of the material, but also complements the professional design.

The living room sets off the foyer with its great momentum, where beige marble overspread around to highlight the precise ratio of points, lines and

faces with the hue of calm and staidness. Along the turning and twisting of lines, British language of elegance and grace comes. Out of the furniture, textures go in an ultimately classic vein and grain with the aid of lighting far and near, weaving a typical luxurious image that's exclusive to the British classical dignity and elegance. Baker chairs, warm and classical, as well as crystal chandelier above, inject beauty and refined quality. Between the kitchen and the dining room, stands simple bar counter to partition, both of which are in a similar or exactly same hue to present coherence and a flowing vision.

The master bedroom to continue the classical sense of the public space is designed into purity of white, just to respond to the British classical. Material

conversion is remarkably to define spaces public and private. Rosewood flooring and white carved walls undoubtedly contribute to a sweet sleeping. Out of the main wall coated in elegant fabric, and bedside light of symmetry, are spatial layers lit up by wall lamps to highlight grace and amenity in a classical air of peace and quiet. With the standing of a hollowed-out screen between the sleeping area and the sitting room, narrowness is then grounded off to create a space somewhat transparent. The bathroom affiliated with the master bedroom is destined to a summit of luxury. The whole space seems to be cut off a whole marble piece. Marble strips and bars, as well as size-enlarged wash stand with a good storing function, give off an air that's really an icy gem to the British classical.

馥曼优雅，法式古典殿堂
French Classical Palace

设计公司：达特思室内设计（北京）有限公司
设计师：翁伟锴
面积：1 050 m²
主要建材：天然大理石、手扫漆木饰面、金箔线条

  本案以"法式新古典"为主题，将古典与现在相结合，以简洁明快的设计风格为主调，在总体布局方面尽量满足业主生活上的需求。主要装修材料为手扫漆木饰面，以金箔色线条做装饰，墙面各种雕花线条尽显贵气。

  步入大门，映人眼球的是带有雕花的玄关墙，让人眼前一亮，客厅内稳重大气的壁炉造型墙面与整个空间相呼应。线条的衬托，使整个空间温馨明快，又不失法式情调。

  天花的吊顶以平面为主，没有过多复杂的造型，简洁大方，与整个墙面的造型很好的呼应。

  古典的装修风格摒弃了简约的呆板和单调，也没有古典风格中的繁琐和严肃，让人感觉庄重和恬静，适度的装饰也使家居空间不乏活泼气息，使人在空间中得到精神和身体上的放松，并且紧跟着时尚的步伐，也满足了现代人的"混搭"乐趣！

Design Company: Details Design Consultants
Designer: Weng Weikai
Area: 1,050 m²
Materials: Natural Marble, Sweep Hand Lacquered Finishes, Gold Lines

  This case revolves around the "French neo-classical" as its theme, classical and now the combination of the main tune, concise design style, try to meet the needs of the owners of their daily lives in terms of overall layout. The main decoration materials for hand sweeping lacquered finishes, gold color lines, decorative wall a variety of carved lines with a full extravagance,

  Entered the door, suck the eye is the entrance wall with carved, some bright spots, stable atmosphere in the living room fireplace shape of the wall echoes the entire space. Lines of the background, so that the entire space Wenxin Ming fast, yet the French ambience.

  Smallpox ceiling to flat without too many complex shapes, simple and generous, and played a very good echoes the shape of the entire wall.

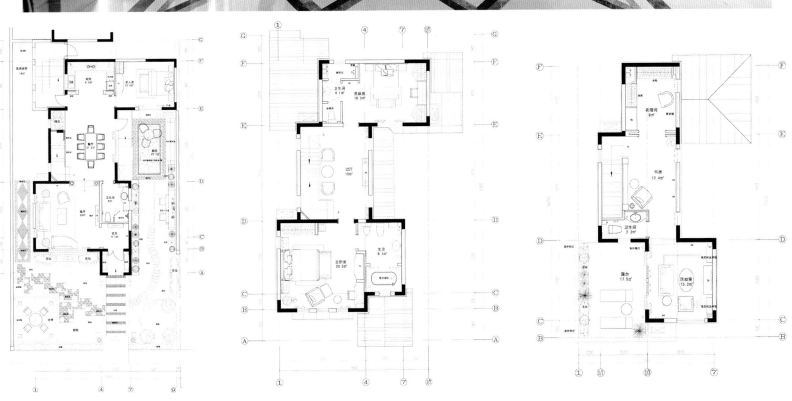

Abandon the classical decoration style simple dull and monotonous, there is no cumbersome and serious in the classical style, people feel solemn and quiet, modest decorative home space there is no lack of lively atmosphere, the people in the mental and physical space relax on, and followed the pace of fashion, but also to meet the modern "mix and match" fun!

安琪尔的梦幻城堡
ngels' Castle

设计公司：达特思室内设计（北京）有限公司
设计师：翁伟锴
面积：550 m²
主要建材：天然大理石、手扫漆木饰面、金箔线条

Design Company: Details Design Consultants
Designer: Weng Weikai
Area: 550 m²
Materials: Natural Marble, Sweep Hand Lacquered Finishes, Gold Lines

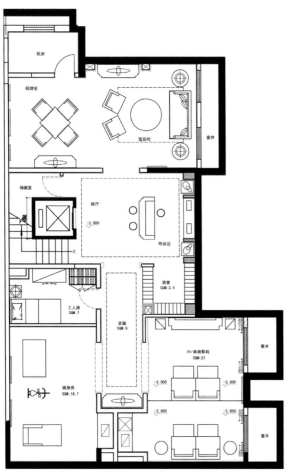
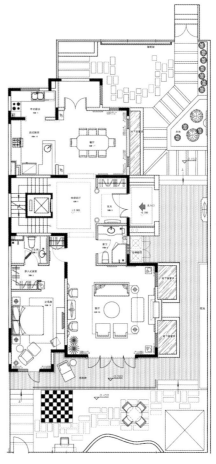

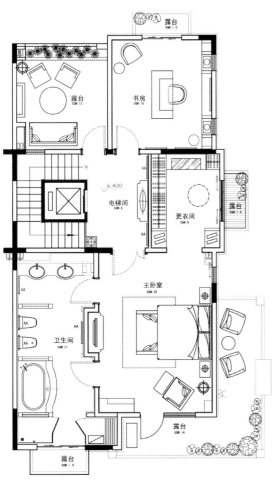
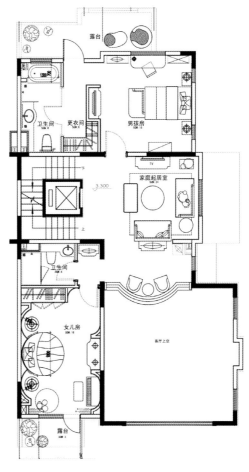

本项目为地下一层加地上三层高别墅样板间，依据甲方发展商的要求创造一个欧式华丽的居住空间。

在主导适合中国人居住的建筑布局设计内，改善并优化原有空间的布局，强调空间的高度感，注入经典欧式新古典及局部混搭的设计理念，主要空间使用大量高档装饰材料及石材，最明显的特征，即是在木饰面上的洗白处理及配色上，均使用华贵欧式植物元素金箔线条，新古典独有的隽永质感线条造型，流露出源于欧洲历史艺术文化的背景。

同时结合软装设计的力量在各个部位上的摆设搭配，家具、灯具、布艺、装饰品、饰品的精心挑选，以及新古典家具的线条、纹饰造型经过工匠的重新诠释，突出华丽感及舒适度，营造出适合中国人居住的欧式豪宅，品味生活的舒适空间。

This project for the basement and the ground three-storey villa model, in accordance with the requirements of the Party developers to create a European gorgeous living space.

Dominant building suitable for Chinese people living within the layout design, improve and optimize the layout of the original space, emphasizing the space a sense of the height, injected into a classic European-style neo-classical and local mix and match design concepts, the main space using a large number of high-end decorative materials and stone. the most obvious feature, even if washed processing and color of the trim surface are luxurious European plant element gold lines, neo-classical and unique timeless texture of the line shape, showing the background originating in the European history of art and culture.

Combined with the strength of the soft loading design - in all parts of the furnishings with; carefully selected furniture, lamps, fabric, decorations, depending on the goods, as well as the lines of the neo-classical furniture, decoration and shape after the re-interpretation of the craftsmen, highlight the gorgeous sense of comfort, creating a European-style mansion, suitable for Chinese people living in the taste of life comfortable space.

欧洲经典，贵族格调
The Classic European, the Style Noble

设计公司：LSDCASA
设计师：葛亚曦
面积：655 m²

Design Company: LSDCASA
Designer: Ge Yaxi
Area: 655 m²

这是一个被混搭得很个性化的家，既可以看到融合了法国、意大利等欧洲经典的格调，又能够找到清代中期的中式味道。通过精致而丰富的细节，LSDCASA为整个空间带来漫无边际的贵族气息。

此套别墅位于北京门头沟，属于典型的山地别墅，拥有完美的自然资源和景观体系。如何通过软装设计放大项目的核心价值，同时为一种已成经典的风格体系带来设计的新意，是着手设计之前，摆在LSDCASA团队眼前的问题。坚持不走软装设计界寻常路的LSDCASA，再次将出色的设计融入空间中，以一种崭新的生活方式诠释豪宅的真正意义，他们将此次的住宅风格定义为经典欧洲混搭泛东方文化。

进门处的双玄关，是对整套别墅风格的一个完美提炼。精致的巴洛克玄关台，搭配的却是一幅中式的挂画、花瓶、镜子等小件，全部采用了中瓷西做的手法，蓝白相间的青花瓷，其清秀的姿态让空间变得含蓄起来。

A mashup this space this is, where styles of France, Italy and Ming and Qing are expressed in details rich and exquisite to allow for a noble sense that spreads beyond.

The location of Mentougou, Beijing, provides a typical mountain villa to enjoy a perfect natural resource and landscape system. How to enlarge the core value by upholstering and make innovation through a classic style poses to be a challenge. Nothing can defeat LSDCASA, as a team who sticks to unique design and makes it into space. And this project, beneath their skillful hands, is positioned a combination of European classic and eastern culture.

The double entrances are actually a perfect refined episode, where the baroque stairs are fine but equipped with a Chinese painting, vases and

起居室的正厅，硬朗的皮革、石材及兽纹营造出了男性的味道。为了不让空间显得过于拘谨和庄重，花艺以及蓝色色块的注入起到了非常大的作用，同时壁炉上方那幅有点调侃意味的挂画——一个滑冰的牧师，更是为空间带来了一抹轻松和诙谐。与此对应的偏厅，一把古琴、一幅中式大挂画，则变幻出女主人的温婉与涵养。

在极具仪式感的7米挑高的餐厅中，马蹄形的拱窗上整面墙挂的窗帘，增强了空间大挑高的视觉效果。餐桌椅以及餐具无论是造型、漆色还是摆放方式，都完全重现了欧洲文艺复兴时期的西餐标准仪式。墙上的那幅中式的金帛挂画，在色调上与空间中的物品达成了很好的呼应。LSDCASA通过从细节到整体的微妙处理，为空间带来无限的灵性与贵气。

整个二层是女主人和女儿生活的空间。主人房的灰粉色调，柔美而不失品位，洛可可时期对女性文化的推崇在这里得到了充分的展现。家具上的拼花、手绘及雕刻再一次重现了文艺复兴鼎盛时期对家具制作的苛求。三层空间则全部为男主人所有，作为一位严谨、成功的金融家，其对细节的苛求，也在这里得到了完美的诠释。

在地下一层的家庭阅读厅中，LSDCASA完成了一次新乔治亚风格的经典再现，为了强调出文化符号的特征，这里的四面墙都做成了书柜，触目所及、触手可得的全是书。色彩浓重的各种布艺，搭配"体态臃肿"的沙发，让大家在这里感受到更多的是一份舒适与随心所欲。偏厅的小阅读区，则散发着一丝女性的柔美与隽秀。

mirrors. All Chinese elements are treated in western approaches. And the blue and white porcelain instantly leads the space into a delicate gesture.

In the hall, tough leather, stone and animal pattern all together bring forth a masculine atmosphere, the solemn and formal of which is grounded off by floral and blue patches. The painting above the fireplace of a skating pastor looks funny and interesting. In the side hall, from the seven-stringed plunked instrument, and the large painting of Chinese style, sprouts out the hostess's conservation.

The 7-meter-tall dining room is very ritual. The arch window is horseshoe-shaped, which with the curtain, enhances the visual broadness. Dining tables and chairs and tableware, as for shape, paint and gesture, completely reproduce the standards of Western food during Renaissance. The golden silk picture on the wall well corresponds to the items in terms of color. Such handled details and entirety endows the space with infinite spirit and extravagance.

The entire 2nd floor is for the hostess and the daughter. The master bedroom is soft but without losing quality, where the esteem for female of the Rococo period has been fully demonstrated. The parquet, the hand-painting and the sculpture once again reproduce the excessive demands for furniture during Renaissance. The 3rd floor for the host, a rigorous, successful financier, makes a perfect interpretation for demanding details.

The reading hall in the basement accomplishes a new Georgian-style classic reproduction, where in order to emphasize the characteristics of cultural symbols, four walls are made into bookcase, and fabrics of all kinds, are matched with "bloated" sofas, so that comfortable and doing at wishes can be available. Meanwhile, the reading area in the side hall exudes a hint of female softness.

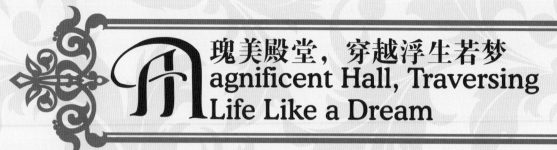

瑰美殿堂，穿越浮生若梦
Magnificent Hall, Traversing Life Like a Dream

设计公司：深圳金凤凰装饰
摄影师：江宁
面积：1 050 m²

Design Company: Shenzhen Golden Phoenix Decoration
Photographer: Jiang Ning
Area: 1,050 m²

本案由心家泊地产开发，占地面积达1 050平方米。开阔的面积，使得空间展现出磅礴大度的雄伟气势。进入室内，即对客厅之琉璃艺术的印象深刻不已。以自然垂落于客厅中央的水晶吊灯为例，其姿态典雅稳重、晶莹冰清，烘托出空间的利落大器，辉煌一片。立面以有着丰富图腾的琉璃墙体隔断，背景釉色黄中带绿，绚烂夺目，美丽动人。客厅软装主要以优质建材进行精心打造，仿欧式巴洛克家私舒适华丽，软垫与木框匹配，雕工精细，塑造出交际重心的奢华豪宅质感。窗帘的褶皱与束卷色彩柔软饱满，带出整幅窗帘的生气，亦化解内帘硬冷质感与外帘垂感交织产生的厚重感。方正的格局，以图案鲜明的大理石材铺陈，立柱线脚厚重挺括，与建筑气质融合呼应，搭配质感地毯，诠释出一派新古典意象，仿佛置身于欧洲皇室宫廷。

而不得不指出的是，居室的天花大大地绽放出仿古的韵味。挑空的方格设计，外有木纹漆肌理，内嵌银箔包裹的繁复肌理，凹凸有致，内外有别，加之线条压边得当，与充满现代感的吸顶灯相匹配，衍生出一种前所未有的魅力风华。

在餐厅的规划上，设计师以采用欧洲式的餐厅设计，运用大量的木质材料作为空间主调，从天花到餐桌摆置，无不隐约透露着时尚与古典互为交融的自然古朴风韵。餐桌正面方向设有壁炉，乃石膏灰制作，浮雕艺术了得，一丝一毫皆可体现出设计者的匠心独运。此外，立面开窗引入的自然光线与人工光源彼此呼应，洒落空间的每个角落，营造出微醺的曼妙氛围，扣人心弦。

卧室大尺度的空间，天花采用开敞式设计，中部如天井般挑空，以轻纱薄幕加以遮掩，若隐若现的间接美感，唯美浪漫。四周实木包裹银箔雕花，展现出王者风范。背景墙则运用花卉纹理的金色帷幕作装饰，华贵古典，更添艺术气息。

这一切的一切都来源于于安妮女王式的建筑风格，其注重细部装饰的程度之深可见一斑，每个局部每个细节均环环相扣，绝无忽略半点空间，可谓内外兼顾。

负一层夹层平面图

二层平面图

负一层平面图

三层平面图

一层平面图

四层平面图

Case by heart home park real estate development, an area of 1,050 square meters. Open area, making the space to show the magnificence of boundless generosity. Enter the room, the living room of glass art impressed endless. Naturally drawn to, for example, crystal chandeliers in the living room the center of its elegant posture steady crystal Bingqing, express a space neat amplifier, a brilliant one. The facade has a wealth of totem glass wall partition, background glaze yellow with green, dazzling and beautiful. Living room soft fitted high quality building materials carefully crafted, European-style Baroque furniture comfortable and beautiful, upholstered wooden frame match, fine carving, shaping the communicative focus of the luxurious mansion texture. The curtain folds the beam volume color is soft and full, with a whole lot of curtains angry, but also to resolve the hard cold texture of the inside curtain and outer curtain drape intertwined heavy. Founder pattern, a distinctive pattern of marble material lay out the column line of feet thick and crisp, echo the integration with the architectural qualities, with the texture of the carpet, the interpretation of the pack of new classical image, like being in the European royal court.

Had pointed out that yes, the bedroom ceilings to bloom antique charm. Pick an empty box design, external wood grain paint texture, embedded silver foil wrapped complicated texture, convex, differentiated, combined with line pressure side proper, contemporary ceiling lamps, and derive a unprecedented charm of elegance.

In the planning of a restaurant, designers to adopt European-style restaurant design, the use of a large number of wood materials as the main theme of space from the ceiling to table placement, without exception, passing along the quaint charm of fashion and classical mutually blend of natural. Table a positive direction with a fireplace, is plaster ash production, relief art terrible, shred can be reflect the designer's imaginative. In addition, the facade fenestration to the introduction of natural light and artificial light echo in every corner of the floating down the space, creating a graceful atmosphere of a little drunk, exciting.

Bedroom, large-scale space, ceilings with open design, the central courtyard like to pick an empty, to veil a veil to cover looming indirect beauty, beautiful and romantic. Surrounding wood wrapped in silver foil carved, showing the manner of the king. The backdrop is for decorative use of flowers texture gold curtain, luxury and classic gift art.

This all comes from the architectural style of the Queen Anne style deep focus on the extent of the decorative detail is evident on every detail of each local interlocking absolutely ignore the slightest space can be described as both inside and outside.

雍荣华府，世袭尊贵
Be Noble and Glorious and to Be

设计公司：林庆宗设计事务所
设计师：林庆宗
面积：1 050 m²

Design Company: Toga Design
Designer: Lin Qingzong
Area: 1,050 m²

本案是中国风与欧洲风完美结合的典范，其唯美细致的风貌洋溢出雍容华贵的氛围，令人神迷心醉。甫入室内，弥漫着原始自然气息的空间跃入眼帘，一室生意盎然的气氛萦绕不断。放眼望去，空间中触目皆是橡木的踪影，设计师极尽描摹居住环境与大自然紧密结合的绿色理念可见一斑。广铺客厅墙体的橡木墙面、实木曲线楼梯、橡木地板过道、层次的天花衍梁、色彩鲜艳的编织地毯与满室微醺的气氛，造就了一个温馨而精致的汪公馆。

客厅家具配置标新立异，欧式布艺沙发组合经典中国风的实木雕刻图腾矮桌，两个国度间的古典元素相互混搭，出乎意料地不显得突兀与咄咄逼人，反而有点俏皮的亲和，且保持了适当的空间感。而最引人瞩目的还是那融入古希腊设计元素的橡木柱式，古典的隽永质感，平淡的外表，处处散发着大自然的气息与浓郁的欧洲文化韵味。为了调和客厅因大量使用橡木建材而造成的光线暗沉的问题，设计者于橡木墙面的正对面高明地规划了大立面开窗，让自然光线能直接穿过窗户进入室内，使得室内光线充足。

为了将新古典主义风格进行到底，设计师对于汪公馆卧室的设计也非常考究。极具明清时期家具意味的架子床吸收西方的多种工艺美，雕刻、线形简单古朴，甚有历史味道，因而成了整个卧室的焦点。欧风乡村沙发与地毯徜徉在卧室之中，温情脉脉，与厚重深色的橡木建材交织在一起，把田园风情演绎得淋漓尽致，带给人一种自然闲适的感觉。

A paragon the project is, of Chinese and European styles, where the aesthetic and meticulous reveals an elegant and enchanted atmosphere, but primitive, and vigorous. The oak materials employed in large amounts in walls of the living room, the curved stairs, the flooring of the aisle, and the girder of the ceiling, confide in a green idea of the living environment closely combined with nature. When woven rugs come colorful, the wooden texture makes a warm and delicate mansion.

The furniture in the living room is unconventional: classical elements of European fabric sofa and Chinese style wood carving are unexpected but not towering, or unobtrusive. Instead, some affinity is implied within while proper spatial sense is maintained. The most eye-catching is the oak pillar, an ancient Greece element with a classical timeless texture and an ordinary appearance that always exudes an atmosphere of nature but abundant in European flavor. Then landing windows is timely to take daylight that neutralizes the stiffness and the dark due to the large-scale application of oak.

The bedroom continues the neo-classical style, where the canopy bed takes in western variety of techniques, and becomes the focus there with sculpture and line simple, quaint and historical. On the carpet lies sofa of European village, which together with oak materials, brings forward the idyllic to the full when providing a natural feeling of leisure.

法式浪漫，精巧秀丽
French Romantic, Delicate and Beautiful

设计公司：林庆宗设计事务所
设计师：林庆宗
面积：1 050 m²

Design Company: Toga Design
Designer: Lin Qingzong
Area: 1,050 m²

　　中国人对欧洲风情的憧憬是毋庸置疑的，而对于欧洲皇室风情所带来的金碧辉煌更是心驰神往。正因为深谙此道，设计师决意把本案打造成一个荡漾无限欧洲皇室风情的优质居住空间，让业主享这万种风情所带来的感官震撼，感受华丽贵族所独享的尊贵与品味。

　　进入屋内，触目皆是流光四溢的诱感。整体空间注重色调的搭配，以求营造出浪漫醉人，温情脉脉的氤氲气氛。客厅作为一室之交际重心，设计师费尽心思，在软硬装饰上可谓精益求精。造型优美且舒适的家具徜徉在客厅的中心地带，纤巧的浮雕艺术加上温润如玉的质感使其宛如一个精美绝伦的工艺品，供人鉴赏。在微醺的灯光映射下，让人有想轻轻抚摸的冲动。为了形成一个完整的室内设计的新概念，设计师特别选用了跟屋内家具的形式完全一致的墙壁装饰和室内陈设。优美的曲线框架、织锦缎的欧式窗帘、表面镀金的装饰品以及高贵典雅的地毯，点滴奢华得到自然的流露，空前完美。餐厅与客厅连成一气，延续了客厅的大气风华。餐厅大量使用橡木建材，大至墙面柱式，小至柜台桌椅，与大理石铺陈的地面辉映成趣。此外，餐厅的独到之处在于规划了中国明清古典红木圆桌，弦纹明晰，色泽亮丽，携手氛围浓郁的欧洲格调，演绎了独特的混搭传奇。

　　推开房门，卧室墙面以典雅的壁画悠扬渲染，绘工雅致，灵动自若，犹如一幅水墨般的画卷在眼前徐徐展开。有别于客餐厅的富丽堂皇，卧室主要以浅色调为主旋律，银灰色的浮雕化妆台与靠椅组合同色系的云线连头窗帘，线条勾勒精致，光泽、柔和、圆润之余略带丝屡时尚感，容易让人产生仿佛置身于华丽宫廷的错觉。

There is no doubt that Chinese tend to have a good preference for a European style, particularly the royal magnificence, which requires a space of this project where to allow for occupants to enjoy amorous feelings brought about by sensory shock, dignity and gorgeous taste exclusive to the aristocracy.

Everywhere strikes with an impression of overflowing temptation. The overall space focuses on the tone with a view to creating a romantic intoxicating and sentimental. The living room paid painstaking significance on features a special upholstering and soft decorating. Elegant and comfortable furniture sits in the center. Relief with jade texture certainly makes it an exquisite handicraft. Wall decoration and furnishings continues feelings of furniture to make an entire interior design: the framework is curved, the curtain of brocade is European, the adornment is gold-plated, and the carpet is noble and refined. The dining room joins the living room and continues its magnificence, where oak materials in amounts go to the construction of walls, pillars, counters and tables and chairs, making a sharp but interesting marble ground. Additionally, a legend of mashup comes into being with circular wooden table of Ming and Qing style.

In the bedroom, walls are rendered with melodious, elegant murals, like an ink scroll painting to be slowly unfolding. Here is space, quite different from the grandeur of the dining room, where the light hue is persistent: embossed dressing table, arm chair group and curtains are of silver, and lines are sketched exquisite, glossy and soft. It's bound to be a place you can get an illusion of being exposed to a palace.

清新甜美，温情脉脉
Fresh, Sweet, Sentimental

项目名称：卡内基山间联排
设计师：汤姆

Project Name: Carnegie Hill Townhouse
Designer: Tom

　　一般说来，大户型居室的业主多为成功人士。对于生活，他们有着高品质的追求，希冀达到"我的空间我做主"的理想境界。然而，在设计师看来，把本案的大户型结构打造成一个内敛大气，与居住者个性文化特质相吻合的生活环境才是王道。

　　对于本案，其优势在于拥有宽敞的空间，有利于设计者进行多区域划分，满足业主多方面的需求。整体空间大至分为前厅、客厅、餐厨区、主卧、男孩房、女孩房、工人房等几大板块，各板块间相互连接、呼应，流线完美，通畅无阻。

　　软装方面，无论从色彩、材质抑或线条等方面均体现了设计师的严密思维及细腻心思。如家庭居室，相对于其他板块以或简约或繁复的墙面挂画的形式来装修墙体，

家庭居室的墙面设计皆以精致的细条竖纹进行勾勒，搭配颜色绚丽的横条斑纹地毯，动感充盈，有利于拉升空间视觉。纯白色的背景上嵌入壁龛以作收纳电视之用，这样的规划使得平板电视与墙体融为一体，实用之余亦保证了墙体的完整美观。布艺家具颜色搭配掌控有度，深啡、淡黄的沙发依偎着土黄的长形小矮桌占据家庭居室的中央地带，既不过分深沉，也不显得单调。

金漆镶边镜面与矮桌、装饰壁炉、花纹繁复的毛地毯以及如珠帘般垂下的烛台吊灯或台灯，配合木质桌子与编织的收纳篮子等等铺陈整个客厅空间，通过现代的设计手法及材质还原古典气息，把新古典主义"形散而神不散"的精髓发挥得淋漓尽致。同时，设立大面落地观景窗，清新空气自然对流，阳光美景一览无遗。

主卧淡雅清新，以白色为主轴，辅以淡黄色加以描绘。床榻、沙发座椅及窗帘如出一辙，均采用淡黄色雕花纹理的织锦缎为主要材质，以此提升整体空间的典雅精致氛围。绿色，大自然最本质的颜色，清新自然，被设计者运用于对工人房的着墨铺陈，身临其境犹如置身于绿色的海洋，呼吸着甜丝丝的清新空气。女孩房跟男孩房的出彩之处莫过于利用宽阔的空间环境各自配套了格调迥然不同的盥洗室。充满阳刚之气的男孩房整体墙身为天空的颜色，明快纯净，加上多功能的家具组合，让房间更显宁静淡定，颇具安全感。洗漱区则以银灰色墙纸搭配似砖头的瓷砖裸露为主，茶色的竖纹拉帘取代屏门，横亘其间，隔断干湿区域，具有一定的层次感。粉红色作为女孩子闺房永恒的主题，无可避免地被选用为本案女孩房的最佳主色调。从墙纸、床榻、地板，再到盥洗室，无一例外地被应用于整个空间。这个从来只属于童话故事里的公主或是芭比娃娃的颜色，经本案设计师之手俨然成为缔筑现实版公主城堡的主角，让温柔、青春、浪漫、温馨的气息弥漫空间的每一隅。

As is often the case, a dwelling space of large size would be owned by distinguished people. Such people usually have a high-qualified pursuit for life to achieve realization of my space to be mine all around. According to professional designers, however, the shift or transformation of a large space into a living environment reserved and fitting in well with the occupants' cultural property can be truly spoken highly of and worth respect and attention.

This project, a space of large size, is feasible and convenient to be divided into sectors for multi-functional needs. Blocks are interlined, corresponding in a flowing state, of the vestibule, the living room, the kitchen, the master bedroom, rooms for the boy, the girl, and the workers.

A thorough reflection of thinking and wisdom the upholstering is in terms of color, texture or lines. One good example is the drawing room, where compared with walls elsewhere that are decorated with paintings whether simple or complex, walls are outlined with fine lines, which expand the visional effect particularly with colorful grained carpet. Embedded in the white purity, is niche that guarantees the physical beauty of the wall and that flat-panel TV can be fully integrated with wall. Fabric furniture stands varying in colors. And In the center, sofas, dark coffee and yellowish sit next to the small yellowish brown table, neither too deep, nor monotonous.

In the living room, modern design techniques and materials successfully restore a classical atmosphere, of gold lacquer trimmed mirrors and low tables, decorative fireplace, wool carpet of complex pattern, chandeliers, lamps, wooden tables and woven storage baskets, which brings out to the full the essence of neo-classical of forms not unified but with a common spirit. Across the large landscape windows, comes not only the fresh air, but also beautiful views.

The white hue of the master bedroom is complimented with yellow. The bed, the sofa, the chair and the curtain are mainly of yellow brocade, which is bound to enhance the atmosphere refined and elegant. Green, the natural color, is available in the worker's room, which allows for an ocean exposure and fresh air. Rooms for the boy and the girls feature washroom of its own. In the boy's, walls are in sky color, furniture are multi-functioned, the space thus becoming quieter and more tranquil than ever, and the washroom is decorated with silver gray wall paper, brick-like tiles, tawny stripped curtain instead of door to separate the dry and the wet. The girl's is themed as purple that spreads throughout from wallpaper, to bed, to ground and to the washroom. Thanks to the skillful hands, a color that's can only belong to a princess from a fairy tale or Barbie, it here serves the princess living in reality or in front of your face, to fill with every corner of an air that's gentle, youthful, romantic and cozy.

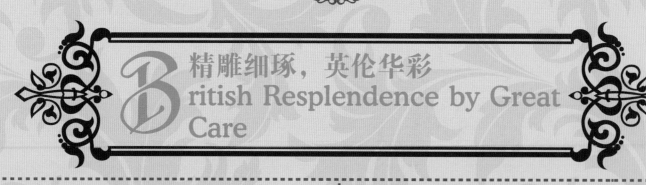

精雕细琢，英伦华彩
British Resplendence by Great Care

设计公司：深圳金凤凰装饰
摄影师：江宁
面积：400 m²

Design Company: Shenzhen Golden Phoenix Decoration
Photographer: Jiang Ning
Area: 400 m²

　　名为安妮女王的本案，伫立于福州五四北观音峰，建筑外观庄重恢弘，细部装饰精雅奢华，把经典的安妮女王式建筑风格描述得惟妙惟肖。浪漫高雅的居室内，如景似画，独具英伦风情的韵味，完美地诠释了名流贵胄的尊贵生活品位。

　　安妮女王为多层别墅，因而十分注重对外部轮廓的刻画。其仿效法国西北部和中部地区的城堡、庄园及农舍的建筑形式，打造似圆锥式的尖塔或其他非对称平立面屋顶，鲜明突出，具有极高的鉴赏价值及宗教色彩。

　　而对于内部装饰，设计师更是秉承精益求精的态度，细心研磨，力求做到尽善尽美。设计师注意使用通透的木质电梯，以求达到消除人与机器的距离，避免让居住者产生生硬冰冷之感。客厅天花和挑空方格设计，内里线条明朗，层层递进，镂空雕刻，如花绽放，光芒外露。地面以纯手工精选的大理石建材为基调的空间色彩搭配，让悬于客厅天花中央的水晶吊灯更显干净透亮。素雅如兰的锦缎沙发雕花纹理通顺，质地紧密细腻，置于其中，与质地丰厚的毛绒地毯相互衬托，繁复之中却不见厚重。一如餐厅的格局安排，徜徉着新古典的典雅风格，宛若一座白色梦幻的浪漫城堡。精美的银器制作工艺精湛，与设计师的创作灵感邂逅，被普遍地陈列于展览柜及茶几上，演绎极致的贵豪生活。考虑到室内层高的安全问题，设计师精选纯白欧式新古典风格的木制雕花玉栅栏作为保护围栏，实用性极强，亦给人留下很好的艺术观感。

　　家庭厨房作为一家人共聚一堂的重要区域，沿袭皇室宫廷的设计格调，成就一片金碧辉煌的宫殿效果，书房亦然。金装包裹的餐椅及柜台，线条勾勒圆润美观，与带灯罩的金漆吊灯相得益彰，华丽璀璨，让空间添色生香，彰显主人的高贵气质。

　　至于私密空间的规划，以纯白色作为居室的基础色，配合紫色花纹背景墙，

丝毫不显单调乏味，反而增添了些许的典雅风情。线条感颇强的白色装饰品置于床头板上端，紧贴着背景墙，起到画龙点睛的作用。向室外凸起的飘窗顶端设多角形的尖塔，既有利于进行大面积的玻璃采光，使得室内空间在视觉上得以延伸；又增强了建筑的表现力。此外，飘窗亦可作为阅读的美妙一隅或观景台。躺在窗台上，或坐或卧，或看书发呆，或与亲朋谈天说地，或昼望车水马龙，夜观满天星斗，好一派温馨静谧的景象。琴房与卧室一脉相承，优雅的白色蕾丝床具，黄绿色的背景墙，花纹生动空灵，给场域增加了无限的动感。

安妮女王别墅是英伦风情与东方智慧彼此融合的结晶，是英伦建筑的经典，身处其中，触目皆是如艺术品般的完美景象，给人以恣意的享受。

Called Queen Anne's case, standing in Fuzhou May Fourth North Guanyin peak, solemn and grand architectural appearance, detailed decoration delicate luxury, classic Queen Anne architectural style is described vividly. Romantic and elegant living room, like painting, such as the King, the unique charm of the British style, the perfect interpretation of the celebrities nobles distinguished taste of life.

Queen Anne multi-storey villa, which attaches great importance to the characterization of the external contour. To emulate the architectural form of the castles, manors and farmhouses of northwest France and central regions, to create like cone style minarets or other asymmetric facade roof, distinct and prominent, with a high appreciation of the value and quite religious.

For the interiors, the designer is adhering to the attitude of excellence, careful grinding, strive to achieve perfection. Designers aware of the use of transparent wooden elevator, in order to achieve the elimination of man and machine distance, to avoid the occupants to produce the effect of the stiff and cold and a sense of The living room ceiling or pick an empty box design, the inside lines clear, progressive layers; or hollow sculpture, flower bloom, the light exposure. Ground space with color, pure hand-selected marble building

materials as a central theme for the central crystal chandelier hanging in the living room ceiling is more clean and translucent. Simple and elegant blue brocade sofa carved texture is smooth, close texture, delicate, and into them, set off each other with a rich texture plush carpet, but there is no complicated among heavy. As the pattern of arrangements for the restaurant, and wandering around the neo-classical and elegant style, just like a white dream of a romantic castle. Exquisite fine silver production process, with the designer's inspiration encounter, generally on display in the display cabinet and coffee table, the interpretation of the ultimate life in your ho. Taking into account the indoor-storey security issues, the designers selected white European neo-classical style wooden carved jade fence as a protective fence, highly practical, and also give the impression that the art of perception.

Important region of the family kitchen as a people come together and follow the design style of the royal court, the achievements of a magnificent palace, the study is no exception. Gold wrapped dining chairs and counter lines outline the rounded appearance, with gold lacquer chandelier lamp shades complement each other, the gorgeous bright space to add color and aroma, highlighting the noble spirit of the master.

As for the private space planning, to pure white as the basis of color of the room, with a purple pattern backdrop, no way was tedious, but added some elegant style. The contours are quite white decorations placed on the headboard the top, close to the backdrop, to play a role in the finishing touch. Set to the windows and the top outdoor raised polygonal minaret, both conducive to the lighting of large areas of glass, making the interior space to be extended in the visual; to enhance building performance. In addition, the windows and can also serve as a wonderful corner of the reading or viewing platform. Lying on the window sill, or sitting or lying; or reading a daze; or chatting with friends and family; or day looking busy, starry night view of the school warm and quiet scene. The piano room and bedroom, the same strain, elegant white lace bedding, yellow and green backdrop, the pattern is vivid and ethereal, unlimited dynamic to the field.

Queen Anne villa is the fruit of the British style and oriental wisdom of their fusion is a classic of the buildings of England, who themselves can be seen everywhere such as works of art like the perfect picture gives wanton enjoyment.

怀想巴黎，女设计师的浪漫情结
The Complex of a Female Designer: A Paris Dream

设计公司：戴勇室内设计师事务所
设计师：戴勇

Design Company: Eric Tai Design Co., Ltd.
Designer: Eric Tai

本案以一位留学法国的女服装设计师为设计蓝本。几年在巴黎的生活，让她的内心渗透着浪漫的独特审美，并怀念着异国的景和人。

法国人的浪漫天性是与生俱来的，而女设计师更钟情于古典法式的浪漫情怀。撩人的夏日舞会，星星点点的烛光，悠扬流淌的音乐，在现代法式宫廷风格的房子里，朋友们一边品着香槟，一边侃侃而谈，一边享受着周末派对带来的轻松和愉悦。

餐厅地面黑白相间的云石，极具法式古典主义的风格特色。宽敞的客厅里，设计师特地设计了一组可以容纳很多人的沙发，让好客的女主人可以在周末尽情地邀请久违了的朋友，叙旧、畅想。橄榄绿的墙面加深了整体的自然和怀旧气氛，欧式的经典造型天花和古老的铜制吊灯，仿佛又回到巴黎古老的建筑里。沙发、餐椅、酒杯、烛台，花艺，所有的家具摆饰都带有优雅的贵族气质和迷人的法式情怀。

女主人的工作室设在地下室的车库，无数的设计创意和作品在这里诞生。这里陈列着主人收集的古老的缝纫机，喜欢的犀牛雕塑，还有主人的作品和设计手稿。

如果你有幸年轻的时候在巴黎生活过，那么此后你一生中不论去哪里她都与你同在，因为巴黎是一席流动的盛宴。

A female fashion designer who has studied abroad in French is regarded as the design blue print of this case. The life in Paris for many years infiltrates romantic and unique appreciation of the beauty into her mind, and she is thinking of scene and people in foreign countries.

The romantic nature of the French is inherent, while the female designer is deeply in love classical French romantic style. With attractive summer ball, dotted candlelight, and melodious music, in the modern house with French palace style, friends taste champagne and at the same time speak with fervor and assurance, enjoying the relaxation and happiness brought by weekend party.

Marbles chequered with black and white on the surface of the kitchen possess the style feature of French classicism. In the spacious living room, designers particularly design a group of sofas that can occupy a lot of people, so that the hospitable female master can invite friends who haven't met for a long time to talk about the old days and to give free

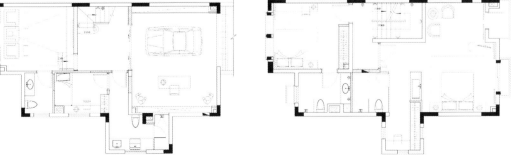

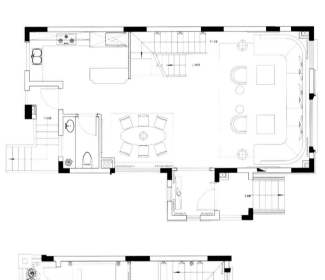
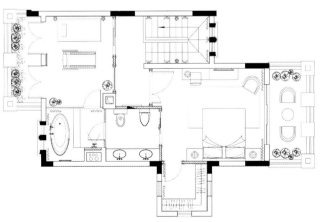

rein to their imagination on weekends. Olivaceous wall surface deepens the overall natural atmosphere, with the ceiling of classical model of European style and ancient droplights made from copper, and it seems as if we have returned the ancient building in Paris. Sofas, dining chairs, wine glasses, candlesticks, and floricultures, elegant noble temperament and charming French style are mixed in all furnish and accessories.

The working room of the mistress is set in the garage of the basement, and countless design creativity and works are born here. Ancient sewing machine, rhinoceros sculptures, works, and design manuscripts are exhibited here.

If you are luck enough to have lived in Paris when you are young, no matter where you go in your future life, she will be with you, because the Paris is a movable feast.

浓情意大利，贵族新风尚
An Italian Flavor, A New Aristocratic Trend

设计公司：香港方黄建筑师事务所
设计师：方峻
面积：690 m²

Design Company: Hong Kong Fong Wang Architects and Associates
Designer: Fang Jun
Area: 690 m²

在这套别墅样板间中，设计师试图诠释出意大利新古典的格调，整个设计以更加温馨的手法营造出引人入胜的视觉效果，使大家与空间有着更多的互动。新古典风格有着更多的意大利18世纪元素来衬托出空间的精神，也有着更多的表现形式。

挑空两层的客厅是整个别墅设计的重点，高大的落地窗，挑高的空间流露出贵族的大家风范，壁炉所在的主背景墙以大面积的金色装饰和超大的大理石框套尽显豪门气派，与紫色高级地毯和黑色真皮沙发共同刻画了高贵的气势。起居空间设计则给人以尊崇感，同时也延续了主要的客厅装饰元素，黑白相间的条纹元素呈现出的是一份既现代而又古典的气韵，中间的装饰与点缀使空间呈现出更加良好的视觉效果。而空间的动线设计让人们走在空间中可以更加随意而自然，不至于受到过多的干扰，各种体验不言而喻。

主人套房整体米色调搭配富于贵族气质的家具，更体现着低调奢华的生活态度，卫浴、衣帽间及生活起居功能布局合理，充分体现别墅的生活尺度。地下室部分则是一个完整的私人会馆，影视房、健身区、桑拿、酒吧等功能齐全。闲暇时，品一杯红酒，倾听着高大的壁炉里噼啪窜动的火苗声，自由徜徉在古典奢华和浪漫雅致之间，尽情享受生命的绚烂与极致。

A villa showroom this space is, aimed to interpret an Italian neo-classical style, warm, and visually appealing to generate people more interaction. And a space it is, where the neo-classical style has been involved more Italian elements of the eighteenth century, and more varying forms.

The living room raised to a double height is fixed with French windows, super-large marble frame, fireplace whose backdrop is decorated with gold ornaments of large areas, purple expensive carpet and black leather sofa. All together bring out the air and manner exclusive to a mansion while contrasting to and echoing with each other. The living space continued the elements used in the parlor, where black and white stripes look both modern and classical, and in the middle of which, the decoration and embellishment compliment the better visual effects. Lines go free and natural, without much disturbance or interference.

The master suite coated in cream color is supplied with furniture that is rich in aristocratic sense, but reflects a low-key attitude towards life. The layout there including the bathroom, and the cloakroom, is reasonable, fully embodying the scale of a villa. The basement is completely facilitated with a private hall, a movie room, a fitness area, a sauna section, and a bar. A red wine by the crackling fireplace, is a stroll through the classical luxury and the romantic elegance. That's an enjoyment gorgeous and ultimate in life.

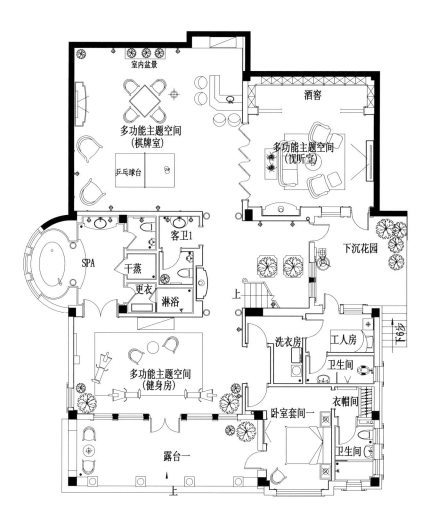

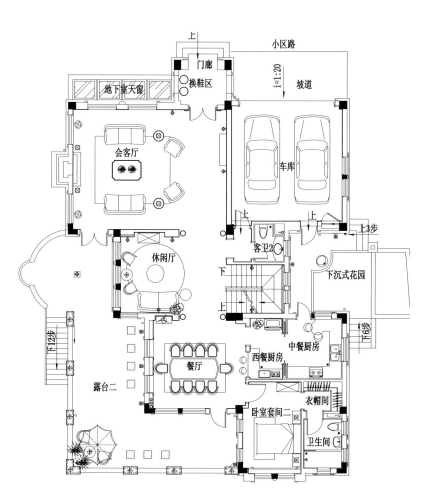
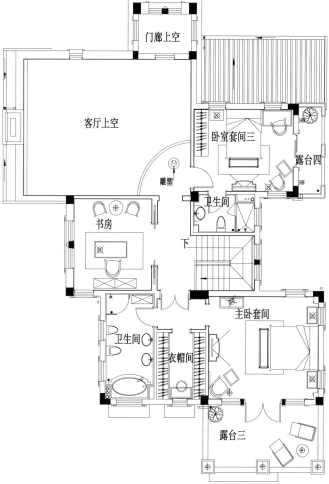

简于形，奢于内，尊享艺术飨宴
Internal Simplicity, External Extravagance

设计公司：卓别林设计师事务所

Design Company: Chaplin Associates

架构没完成之前，空间部署、铺陈就要到位，是本案设计的难点。室内元素的铺就也要在架构未作之前完成，如烟筒、餐厅家具等等无疑使本案难上加难。

经典的背景下，原有元素的利用，并富有创意的新鲜事物，先进的内饰是本案折中的表现。翠蓝色系、赤土色系运用于空间，恰恰是解决问题的基本方法。不锈钢的操作台，玻璃的装饰，艺品陈设更是高科技元素在空间的尽情体现。经典的长沙发椅，添加几个彩色靠枕，完整的混搭风格又多了几分生气，几分情趣。

如此古典折中的设计，即便老年人感觉心神安怡的同时，也能体会其中的尖端与复杂。

Complexity of the project was that we have got in our disposition finished before building with ready planning structure. Ever more, even some interior elements were ready, namely chimney and furniture for dining room.

We set for ourselves a task to create state-of-the-art eclectic interior, based on classic, combining ready elements and new ones in one style. With this purpose we have used rather ballsy color solutions such as combining of turquoise blue and terracotta. Also in the interior it was

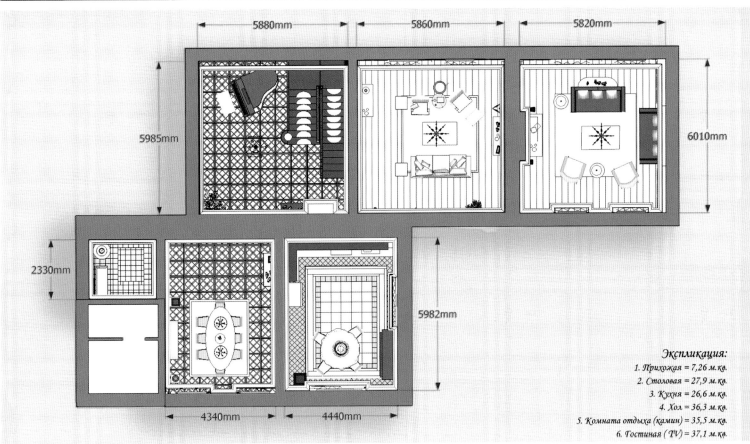

Экспликация:
1. *Прихожая* = 7,26 м.кв.
2. *Столовая* = 27,9 м.кв.
3. *Кухня* = 26,6 м.кв.
4. *Хол* = 36,3 м.кв.
5. *Комната отдыха (камин)* = 35,5 м.кв.
6. *Гостиная (TV)* = 37,1 м.кв.

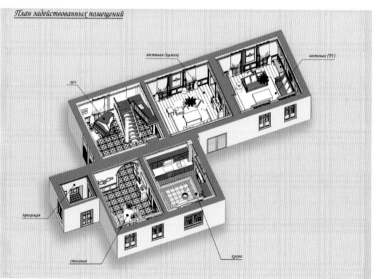

used high-tech elements, namely console tables of stainless steel with glass and art-deco elements. One more interesting element of décor is colored pillows in combining with classic divans which create impression of full stylistic mixture.

As a result we have got eclectic interior in which old age people, focused on classic, will feel them comfortable and sophisticated.

舒雅怡人的生活沙龙
Salon for Daily Comfort

设计师：谢尔盖 | Designer: Sergei

真正的贵族居庭，不在于金堆玉砌的奢侈与繁复，反而更懂得内秀之美，简于形奢于内，由内而外绽放出强大的自信与气场。本案采用简洁的块面分割，精确划分空间比例，藉由端重布局迸发出艺术意象般的视觉飨宴，给人典雅而舒适的体验。

因出自俄罗斯设计师之手，地域文化的差异使得本案呈现出不一样的独特风采，各个功能区都尽可能方正开阔，并且各自独立。本案专门设置两大独立客厅，一间作为承载着全家人欢乐与活动的私密家庭厅，一间则是接待朋友的或满足对外社交需求的公共客厅。而餐厨区也是独立一间，空间开阔裕足，当客厅使用也绰绰有余。三间卧房中，有两房均拥有独立的卫生间与超大的观景阳台，业主可坐在房间里，看风云变幻，时刻享受大自然恩赐的惬意。

至于室内的软装布置，则透露出俄罗斯活泼的民族气质。比如墙壁的修饰，皆采用俄罗斯民间木造房舍广为流传的雕刻母体——花草点缀墙面，造型繁复、灵活别致，色彩完美均衡，酿造出舒适随意的生活氛围。加之沙发灯饰等线条严整，造型柔和，雅致大方，完全契合业主追求个性自由的生活方式。

A true mansion can not be made of luxurious accumulation of gold or jade puzzle, but can sprout out of conservative beauty, simple appearance and extravagant internality. A project this space is to be in precisely external division and at an external proportion, where an artistic image of staid layout allows for an elegant and comfortable enjoyable experience.

Beneath the hands of Russian hands, differences geographical and cultural makes this case a unique style: all various functional areas are as square and open as possible while keeping independent. Of the two separate living rooms, one is for family fun and activities, and the other for social activities. The space of the dining room and the kitchen is also independent and large enough to serve as another dining room. Among the three bedrooms, two are endowed with bathrooms and large balconies, where, in peace and quiet, the family can view clouds and stars to enjoy the pleasant gift of nature.

The decoration confides in a lively national temperament. Walls are coated in complex pattern of flowers and plants, and balanced hues to brew a comfortable and casual atmosphere. That's an approach widely used by Russians to build up wooden houses. Lines like those of sofa, and accessories like lights, are soft and generous, perfectly consistent of the pursuit of individual freedom.

CREATIVE NEO CLASSICAL

创意新古典

蝶儿翩翩舞过浪漫梦境
Butterflies Dance

设计公司：天坊室内计划有限公司
设计师：张清平
面积：80 m²

Design Company: Day Square Indoor Planning Co., Ltd.
Designer: Zhang Qingping
Area: 80 m²

　　对于一个只有80平方米的空间，空间配置一旦不适当，整个居家便会变得非常局促拥挤，让业主心生压迫拘谨之感。因此，对于整个空间的规划，设计师别出心裁，高频率巧用锃明发亮的玻璃推拉门作为各区块间的分界线，加上绚烂夺目的水晶吊灯的配合，造成玻璃面对空间的反射，让原本狭窄的空间在视觉上增大不少。

　　置身屋内，环视四周，本案的真正精髓在于设计师心思细腻，把深度艺术融入室内墙体与天花的精心打造之中，有效提升整个空间的高雅精致品位。从开放式的公共区域到私密空间，对于墙体与天花的砌筑，设计师犹如分门别类般将其分区域打造，可谓各具特色。一气呵成的客餐厅尽管在地面空间的构造上"同气连枝"，但对于天花的设计却各有千秋。相较于客厅中似天窗般的多层次天花，餐厅的天花板却是镂空的蝶式造型，犹如蝴蝶一般翩翩起舞，轻盈的舞姿透露出丝丝的可爱与妩媚。为了达到一个整体的效果，设计师对于餐厨之间以及客厅与娱乐休闲区之间的间隔规划也选用了有着蝴蝶造型印花的玻璃推拉门，身临其中，宛似盈千累万的蝴蝶从四面八方飞聚到家一样，漫天飞舞，何其浪漫。延续这浪漫的气氛，中欧合璧的浪漫古典花纹壁纸铺陈空间的整个墙体，从细部的盥洗室到卧室再到餐厅、客厅，无一例外，皆营造了一个典雅的质感空间。蝴蝶飞舞，百花作伴，花言鸟语，俨然一派世外桃源的景象。

　　此外，设计师特别在色调上加以着墨，透过紫色与白色系的相互巧妙调和，浪漫神秘，既为空间注入活泼的表情，赋予空间极强的亲和力，又平衡空间的整体色调，使空间不至于因过分苍白而显得呆滞。

A space not very large would be become very cramped and crowded if allocated inappropriately. Fortunately, the skillful hands frequently employ sliding glass doors for space dividing, which together with dazzling crystal chandeliers are remarkable and surprising to ground off the original narrowness.

The wisdom, intelligence and effort of the designer blend the depth of art into the interior walls and ceilings to effectively enhance the elegant and refined taste of the entire space. Walls and ceilings from the open public area to private space are endowed with personality and individuality. Despite the unity of the ground, the living space and the dining area are distinguished for features on ceiling, the former's being like scuttle while the latter's being hollowed-out butterfly. In order to achieve an overall effect, intervals between the dining area and the kitchen as well as the living sector and the recreation involves in glass sliding door of butterfly. The continuation of the romantic atmosphere is carried out by wallpaper printed with western and Chinese pattern throughout the whole space, from the washroom, to the living area, and the dining area. With butterflies dancing, flowers in blossom, and birds singing, just like paradise scene.

The whole space is coated in hues of whit and purple. Both communicate and reconcile to bring forward a romantic mystery, to light up the space, and to provide a strong affinity while balancing the overall tone and avoiding the dull and sluggish.

低调高雅，黑色诱惑
Elegance Restrained Temptation Black

设计公司：天坊室内计划有限公司
设计师：张清平
面积：400 m²
主要建材：黑云石、米黄石、金龙石、板岩、黑镜、铁件、不锈钢镀钛、订制水晶吊灯、硅酸钙板、钢琴烤漆

Design Company: Day Square Indoor Planning Co.,Ltd.
Designer: Zhang Qingping
Area: 400 m²
Materials: Marble, Slate, Black Mirror, Ironware, Titanized Stainless Steel, Crystal Chandelier, Calcium Silicate Board, Baking Varnish

永远不会落伍的黑色，紧随着时尚的节奏，被设计师以独特的眼光选中，作为底色铺陈整个家居。非凡的质感，造就帝景苑样品房永恒的经典，给人以厚重深沉的感觉。

本案各区块泾渭分明，偌大的空间从玄关开始，至客厅、艺品区，到餐厅、卧室，动线流畅，有条不紊，把简约高雅的格调描述得酣畅淋漓。然而，设计师并未就此罢手，而是在这简约整洁的基底下附以新古典风格的家具用品与之搭配。两种不同风格的相互混搭，把本案置于另一番风味之中。褐色系的组合沙发稳稳地坐落在客厅的最中央，如梨花般雪白的长绒地毯踏在脚下，有漫步在草坪上的感觉，让人体验俏皮的温柔抚摸，放松心情。黑如墨汁的方形矮桌与之相接应，宛若丝滑的浓情巧克力点缀着一般，给你一场神秘醉人的黑色诱惑。此外，把同一色系的泡茶桌规划其中，无疑给整个空间增加了丝缕的禅意。

艺术品区紧挨着客厅而立，展示柜是大大小小的规则或不规则状的格子构造，延续整体空间的色调旋律，同样以黑色为主，林林种种的黑白艺术品与书籍展示其中，熏黄的灯光下，容易让人情不自禁地将它当做博物馆，而忘记这是自家居室的艺品区。这，是浪漫的艺术气氛使然。而客厅则区别于一般居室的设计，不与厨房或客厅相连，反而多设了一个起居室，集用餐与休闲于一体，极具人性化。雕花纹理的餐椅，曲线有致的墨黑餐桌，晶莹璀璨的琉璃餐具与吊灯，彼此衬托照耀，好不瞩目。墙面设计不再以繁复的花纹壁纸加以装饰，取而代之的却是以简单如斑马线状的浅灰色系壁纸粉饰，与光滑的大理石地板互相辉映，低调，不显张扬。卧室空间宽敞，如同波浪般的背墙设计充满动感，顿时让整个空间散发活力四射的魅力。次卧的艺术镂空雕花天花板为居室添色不少，床头上方左右两边的流线水晶吊灯，如珠帘般垂下，梦幻而美丽。

设计如斯，既是时尚的经典，也是经典的时尚。

Black, one color that would never go outdated but always keep up with fashion, is natural to be chosen as a hue for this project, one design, where extraordinary texture, creates a timeless classic that allows for dignity and staidness.

A space it is that's partitioned clearly with the entrance, the living room, the arts and crafts area, the dining room, and the bedroom flanked with lines. A space it is where in a simple and clean setting neo-classical style furniture sits. A space it is where a mashup by two varying styles makes a sharp and intended contrast. In the living middle stands the sofa group of brown hue, while pear-white plush carpet underfoot provides feeling of walking on t lawn. Out of the black square table, sprouts temptation that's like irresistible like black chocolate. And from the black tea table of the same color, grows an air of Zen.

Next to the living room is the area for arts and crafts, where the display cabinets are of different sizes or of lattice regular or not. The same color of black continues the whole tone, while harbors arts pieces and books of black and white. Once light comes dizzy, an image of museum comes into mind, a must result generated by romantic artistic atmosphere. The drawing room different from the general design, is not connected with the kitchen or the living room, but added with a living room. A highly personalized design it is. The chairs are textured of carved pattern, the dining table is black and curved, and the tableware and the droplight are of glass. All are set off, all are contrasting and all are eye-catching. Against the marble ground, the walls, free of any complex decoration, coated in zebra-patterned gray fabric, is reserved and low-key. The master bedroom so broad and airy, looks in vigor and vitality thanks to the wavy walls, while in the secondary bedroom, the ceiling is hollowed out and steam-lined crystal pendent lamps on bed

sides hang dreamy and beautiful like bean curtain.

Such a design makes no doubt classic which is bound to be fashion, fashion that makes another classic.

粗犷圆木，搭建爱的港湾
Love Harbor with Logs

项目名称：罗莎别墅
设计师：煜连、谢尔盖

Project Name: Villa Rosa
Designer: Yu Lian, Sergei

　　本案可谓是设计最成功的作品之一。2010年"乌克兰最佳住宅室内设计奖"，其榜上有名。基辅，莫斯科两年20本专业建筑期刊甚至把其收录出版。

　　室内的开敞空间划分众多功能小区。各区面积虽小，但却极尽舒适。黑、白两色的主题是二元世界的映象，如阴阳之间，生死之间，光明与黑暗之间，过去与将来之间。球形的平面布局到处是此主题的运用。外面暗黑的世界，内里白色明亮。

　　红色激情的运用，极好地消除了黑白二色带来的单调与无味。尽管红色触调与乡村背景的休息气氛不太相融，橘红色的替代却是一种完美、舒适、令人放松的选择。各部件铺陈，如烟囱、织物、门楣、画作、图书、餐具、桌椅等等，在保持独立个性的同时，交织在极具个性的风格里。

　　建筑的立面外形，是对其所在自然风景的补充。人工的建筑，却似乎早已与风、与雨一起走过四季。冰霜的季节，别墅、湖径、喷泉共同成就着一个美好的去处。

　　内里空间，从大厅入口，到阁楼空间，到处弥漫着宏伟、珍厅的气氛。木质板墙、白色粉饰，成水平态势，赋予能量的同时，和谐着、连接着各空间。因为拱腹，客厅因此更加高阔。舒适的休息室，白色的椅子靠近烟囱。如此轻松的气氛，不至于相聚的时候，有些朋友百无聊赖。各样的灯具、枝状的烛台是照亮空间浪漫的星星。晨曦的那一缕光明，穿过大大的窗，洒下一地的舒适。

　　阁楼的利用更是别出心裁，华丽地转身，成了客人房的所在。靠近窗户的两间卧床成对称分布，但中间却夹以木质的隔断。这当然是有效的利用，也是舒适的创造。

　　卧室的空间，自有串连。原是带有甘菊图案的墙纸及幕帘，彼此之间，如同簇拥，穿过窗户的依然是束束光线。

　　厨房宽大，舒适。这里不仅是食物烹制的空间，也是朋友相逢的宴席之地。有了黑色的直纹墙纸、华美光照，不想优雅也难！

　　卫浴极宜放松。浴缸、沐浴、桑拿相互分离。奢华的享受尽在淡淡的色调之间。这里有最先进的铺陈，这里有乌克兰浓郁的民族风情，这里有灯光，这里有舒适，这里当然是一个最完美的设计个案。

205

This is one of the most successful our objects. Not even it was awarded with first premium of Inter Year 2010 competition as "The best residential interior of Ukraine", nor even it was published in twenty architectural slick magazines of Kyiv and Moscow.

First task was the goal to metamorphose cutting house into small comfy palaces. Black and white theme of the project is a symbol of dualism (Yin and Yan, life and death, light and darkness, past and future). It is interesting that in global plan a dualistic theme can be seen in the following: exterior of the house is made in dark color, and interior – in a white one.

The point of black and white style is that it looks like too sterile. Therefore an ideal neighbor which helps to avoid sterile dump is red color. But for principal designer of the project it appears that such solution is too aggressive for country relax. However accents of carroty are a perfect, comfortable and relaxing alternative. Most of décor elements were developed individually. There are chimneys, textiles, doors, paintings and photos appearance. Everything in this project intertwisted in the unique stylistic conception – from color of front to selection of plates and tableware.

Exterior of villa is harmonically completed by general landscape. You get the impression that villa has been here all-time. Summer house and alley to the lake above-water ground with fountain in place are perfect places for rest in frost-free season.

Interior is dominate, vast and recherché. This feeling fills every room, beginning from hall entrance and finishing with attic. Walls are made of horizontally laid on wood, painted in white. This accentuates harmony and connection between all rooms of building, fills space with a single energy. Floor is dark parquet. Soffit is symmetrically lied light beams. They complete the walls perfect. Parlor is much wide thanks to high soffit. Comfortable lounge composition is combined with white chairs near chimney, thus during friend evenings everyone has things to do. Plenty of lamps and candelabrums help to fill space with country romantic. And morning shine through wide windows curtained with light material tenderly wraps the space with coziness.

It is interesting solution for attic: it was transformed into room for guests. Keeping an absolute symmetry it was decided to situate two beds near windows divided with a wood partition. It came very effective and cozy. At such room guests will make themselves comfortable for sure.

In bedroom it is attracted attention by a tandem of creatively different wallpapers with camomiles and curtains. They look like flirting with each other, when light beams through the window.

Kitchen and dining room are wide and comfortable. Here there is place for tasty food preparation as well as for banquet for a lot of friends. Elegancy of kitchen is underlined with dark wallpapers in vertical strips and coziness of kitchen with light.

Bathroom is a perfect place for relax. Here there are bathroom and shower room and sauna. Design of this room is completed principally in light tones. Villa Rosa is perfect example of state-of-the-art ethno design. This building is filled with light, coziness and real Ukrainian atmosphere.

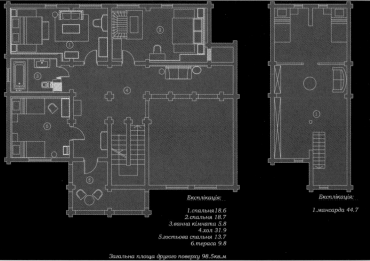

Експлікація:

1. спальня 18.6
2. спальня 18.7
3. ванна кімната 5.8
4. хол 31.9
5. гостьова спальня 13.7
6. тераса 9.8

Загальна площа другого поверху 98.5кв.м

Експлікація:

1. мансарда 44.7

План з розташуванням меблів першого поверху

Експлікація:

1. хол (вітальня, їдальня, кухня) 72.8
2. хол 10.9
3. котельня 9
4. сауна 5.4
5. душова з санвузлом 5.7
6. гостьовий санвузол 1.9
7. кабінет 9.6
8. кімната відпочинку 14.2
9. сходова клітка 7.5
10. тераса 21.7
11. тераса 22.4

Загальна площа першого поверху 181.1кв.м

原木，原乡，原生态
Crud Wood, Native Land and the Original Ecology

项目名称：阳光下的山间小屋
设计师：煜连、谢尔盖

Project Name: Ray of Sunshine
Designer: Yu Lian, Sergei

喀尔巴阡山脉，有很多牧人小屋。本案风格，依然是牧人小屋定位，但风景却相异。乌克兰登山家独特的文化注定本案要完美的气氛，表达着山地小屋新颖质感。外表虽然粗糙、粗制，但却有着优雅的装饰，与不拘一格的织物元素。

无法比拟的冠状造型，有着瑞士牧人小屋的别致，也有着美国小屋的古典。遥远质感的创造，却富有着贵族般的情怀与心胸。

漆黑的木质元素，是内里装饰的极好映衬。手工锻造的装饰，美式的沙发，加长的烟囱，如若分隔开来，必定个性十足。西班牙的瓷砖，是加泰罗尼亚的风情，却有着本地的优雅。众多定制的元素，终至叠加成山间小屋的完美气场。

空间用材极尽奢华，却又极尽精简。相似的空间内，有高端的家具，有"宜家"的地毯。风格原来可以如此 hold。

填充的各种布偶，似杂乱，但实是有序。大片留白的空间，于西方，于乌克兰东部地区，都是经久不衰。漆黑的木质纹理，于留白的背景下，是别致的新颖。

本土文化，但却不突兀。现代的办公家具，一旦主打民间传统的手笔，当又是"有型有色"。

良好的建筑设计，是创意，是生活的真实，是建筑于生活的完美体现，也是专业设计团队辛勤的付出。

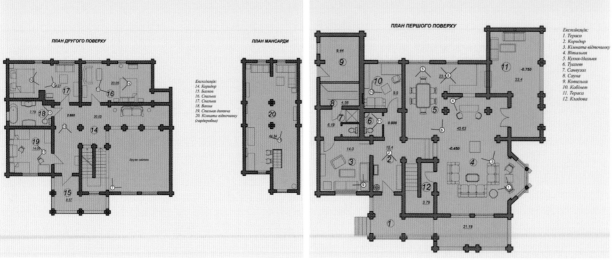

As stylistic solution of the house we managed to convey new vision on Carpathian chalet. Based on unique culture of Ukrainian mountaineers – "hutsuls" – it has been conveyed a perfect atmosphere of mountain shack. But this villa is not just rough, unshaped wood forms, here there are elegant and eclectic elements of décor are weaved as well.

We have made a decision to gather inimitable coronal of all the best mountain styles – Swiss chalet, "hutsul" elements, American neoclassic. It was first task we've set ourselves. A second task consisted in creating a history of the house and rendering an aristocratic to it.

Dark wood is a perfect contrastive background for decorative elements of the interior. And here there are lots of original details. Blacksmithing with "hutsul" ornaments, American sofas and king sized chimney if separately they looks absolutely different. A Spanish tile, which is characteristic for Catalonia, is excellent combined with national dresses. A number of stuff were custom made and together they create perfect atmosphere of a mountain shack.

Experiment also was present in design of interior. Authors of the project decided to combine costly elements with cheap ones. And they came up trumps. High-end furniture and Ikea carpets as it turned out, could excellent live within similar rooms. The main thing is to hold style.

For the first sight it could appears there are too much stuff in the house. Plates, stuffed animals and birds, dishes…you can find here a lot! But decorative elements do not "crush" each other but conversely play with motives. To be closer to folk interior, white painting balks blueprint was used. Earlier it was very popular in Western and Eastern Ukraine. Combining of dark wood and blueprint looks very novel.

In overall picture of the villa there is no obtrusive national culture. Contemporary view of architectural bureau on ethnics evidences that folk motives are able and should be stylish yet again.

Our aim is a sound architectural design. And realization of creative ideas and putting them into life are provided by everyday hard work of team of professionals, cultivating perfectionism in life.

倾世蓝宝石，绝艳维多利亚
Wondrous Sapphire, Amazing Victoria

设计公司：玄武设计群
设计师：黄书恒、欧阳毅、陈佑如
面积：200.97 m²
主要建材：雪白银狐石材、白水晶石材、米洞石、金箔、银箔、水晶、金镜、明镜、图腾雕花版、雪白银狐石材贝壳版、BISAZZA 马赛克、VIVA 砖

Design Company: Sherwood Design Group
Designer: Huang Shuheng, Ouyang Yi, Chen Youru
Area: 200.97 m²
Materials: White Silver Fox, White Crystal Stone, Travertine, Gold Foil, Silver Foil, Crystal, Gold Mirror, Mirror, Totem Carved Edition, Shell Version, BISAZZA Mosaic, VIVA Brick

　　本户为同栋四种房型中次大坪数者，以色彩柔和的"维多利亚风格"来诠释，具有放大空间的视觉效果，更展现此绝世豪宅的精致典雅。

　　由于上海近年来跃升为国际级城市，而随着中心城区可供开发土地的日益减少，供不应求的趋势越发明显，规划优异的豪宅更是一屋难求。远中风华区位十分优越，地处静安区核心区域，距离南京西路商务区仅两个街区。闹中取静，地段稀有，规划全备，加上远雄集团建筑品质的超凡优越，让此区的其他住宅更远远无法望其项背。身为这样傲视群伦豪宅的拥有者，2006 年 5 月开盘的远中风华园现已推出 4 幢共 292 套房源，从售价上看，今年以来报出的 35 000 元 / 平方米的均价，已树立了静安区的价格标杆。对于生活品味的渴望，更高于一般凡夫俗子的追求。

　　生活品味的两个重要元素，一个是价值鉴赏力，另一个则是风格生活实践力。玄武设计在此户中采用维多利亚风格的设计，固然是因为其装饰元素在艺术领域中影响深远，更因为此风格对美学与品味的提升，恰恰切合新上海蓬勃起飞所孕育的新价值。它的用色大胆绚丽、对比强烈，中性色与褐色、金色结合突出了豪华和大器；它的造型细腻、空间分割精巧、层次丰富、装饰美与自然美完美结合，更是唯美主义的真实体现。也因此，维多利亚风格至今仍是许多设计创意元素的来源，更是五星级酒店和壮阔庄园豪宅常采用的优雅典范。

　　在远中风华 SH3-8 大户中，玄武设计运用巧思，撷取 Wedgewood 经典皇家陶瓷艺术美感为设计基调，将维多利亚风格的生活美学充分体现出来。

在空间细部表现上，舍弃了金碧辉煌的过度装饰，塑造一种从容不迫、细致优雅的贵族品味。在白色基底中，渲染柔和的色彩：淡淡的蓝、绿及米黄，如同Wedgewood著名的玉石浮雕般剔透细致，更强调了空间的贵气与立体感。

在天花板及壁面上加入线板、镂空窗花等装饰元素，以婉约线条创造典雅气质的空间氛围，同时以织品的柔软介面，如：帘幔、披毯、抱枕、床饰等丝绸及绒布等光泽织品来营造皇室风格。而图案式的地板拼花及雕花窗棂，从浑厚材料中展现出艺术气质，更是时代复古精神的美好缩影。

浮华光影的美丽新境界——典藏皇家风采的公共空间

公共空间的风格定调贵族典雅，它包含了玄关、客餐厅、厨房。整体颜色运用大量的浅米色及白色，搭配淡金、银箔及少数的黑色描边，与优雅的浅天空蓝。推开玄关的浮雕大门，正如跨进了魔衣橱内的美丽新境界一般，将门外世界的尘嚣一并隔绝在外，蓦然进入皇家风采的城堡宫殿中。玄关地板以水刀拼花呈现的美丽字母纹饰，正如皇家家徽一般，标示着主人尊贵的身份地位，也预告了全户维多利亚式的装饰元素与淡雅风格。

公共空间主要部分皆以岗石铺陈地坪，装修的重点在于色彩、纹饰、线条、质感、以及光泽，透过这几项装饰元素的相互映衬搭配，打造出细瓷精品的剔透质感，呈现上海新贵豪宅的风采器度。

玄关

玄武设计运用匠心巧思，在远雄建设FARGLORY中，撷取F及G的英文字母组合，以石材拼花地坪打造皇室家徽的气势，同时也代表建筑业者对于购屋主人的欣喜、肯定与欢迎之忱。

映入眼帘的，则是用现代及传统工艺手法（雷射切割图腾、贴饰银箔、钢琴烤漆框饰、结合板材及玻璃）共同创造的优雅屏风。右手边则是迎宾的专用鞋柜，门片以灰色镜面组合，高雅的色系及订制的门框，诉说着低调奢华的设计用心。

客厅

近六米长的大尺度电视主墙，利用义大利白洞石做为视觉主体，其天然孔隙纹

▲ 主机移位对照天花板图

路，强调深刻厚实的历史感。中间及两侧的壁面由金镜与明镜组合拼花，搭配细银边框，让整片墙面华丽而不俗艳。

沙发背墙以暖灰壁布为基底，用不同漆面处理粗细线板，如画框般将奶油色的壁布装裱其中。在灯光的照射下，更加深了壁饰的立体感。

天花板特地选用黑金色烤漆组合的线板，滚边般加强视觉效果，灿丽雅致的Swarovski水晶吊灯，特别订制国际知名精品沙发，质感与光泽深获富豪名流喜爱，也界定出客厅的华丽浪漫氛围。

餐厅

岗石地板上描绘了英国精致传统瓷器Wedgewood的经典花纹，直径2.1米的圆形拼花图腾，直接点出空间的风格主题。前方主墙用白洞石围塑出一个造型壁炉，下方以蜡烛灯装饰点缀。从餐桌上细瓷精品餐具、茶具、点心盘甚至茶桌摆饰的考究，带出维多利亚式的生活美学，似乎能看见女主人的优雅身影穿梭其间，一场英式下午茶的悠闲时光即将展开。

主墙两侧是两道双开门，用圆形的银色镂空板描出空间的优雅线条，也巧妙的将公共空间与后方的私密空间分开。

厨房

从餐厅往厨房延伸，是同样语汇的壁饰图腾，以及大面的清玻璃门，白天时可将阳台光线引入。

厨房配备顶级订制精品厨具，中岛式的开放规划，让下厨成为每天最愉悦自得的时光。

淡雅若梦的心灵御花园——典雅内敛的私密空间

穿过走道之后，向左、向右各迈入不同范围的私密空间。在此空间中，一脉相承的固然是维多利亚式的皇家风采，但更多的是淡雅细腻的美学讲究，与私人品位的各自挥洒。

私密空间大量运用不同纹路与质感的壁纸、马赛克拼贴、明镜与线板、磁砖石材，为居住其间的每一位家人，打造发挥个人品位的舞台。这里也是庄园豪宅中的为居住其心灵御花园，让人在复古纯净的美好格局中，忘却一日的辛劳。

世界知名设计师Frank de Biasi如是说："设计是创造力和灵感的结合，是要开发一个房间或一个家的潜力。"玄武设计在此宅空间中充分发挥创意与统合力，每一位名流业主看到属于家宅的独特美丽梦想，原来可以透过设计的灵感与巧思，提前实现。

走道

走道介于公共与私密空间的交界处，适度引导行进动线与心情的转折。隐藏在餐厅的两道拉门之后，整片走道贴上香槟色与金箔色交错的方块墙，当餐厅宴客时，走道灯打亮，宾客透过雕花门扇隐约看到后方闪亮的壁饰，编织成一幅绝美对称的室内美丽窗景。

书房／起居室

书房／起居室内用了模矩式的单元壁饰组合墙面，更拥有整面立墙的书架，如同英国传统的书房风貌。不同的是颜色及材质，设计师大量运用白色板材及落地明镜，在镜面与书墙的互相映衬下，似乎又回到了那个拥有爱与美的艺术，调和物质与心灵的维多利亚文学时代。

琴声如梦，书香轻漫，无论是在此阅读著文学女杰乔治艾略特(George Elliot)的小说，或是品味著罗伯布朗宁(Robert Browning)的情诗，都让人不由得萌发思古之幽情。

主卧室

湛蓝花纹壁纸，搭配高腰白色线板环绕整个空间，创造卧室应有的舒适氛围。床背的主墙采用夺目炫丽的孔雀羽毛拼贴壁纸，透过间接灯光照射，反射出多变灿烂的光泽质感。古典大器的King Size床座采用英国进口典藏家具，优雅柔美中带著坚毅线条，象征着男女主人高贵的地位与不凡的性格。

主卧更衣室

将湛蓝色调延伸至更衣室，柜体利用线性骨架而非板面的拼组方式，加上大面贴饰的落地明镜，提高穿透性，更放大了空间的深度及广度。让男女主人的

Armani、Chanel、Hermes等服饰精品都能各从其类，得到巨星般的保管与呵护。

主卧卫浴

全室墙面以马赛克手工拼贴，展现白色渐层图腾以及中国青花瓷蓝色花纹，在西方的优雅中混搭中式的折衷古典美学。建材为白色带有少许灰色纹路的石材，以全片明镜配搭金镜，并以香槟色银箔线板饰框打造出古典的镜中之镜，同样使用精品酒店最顶级的进口卫浴设备，让入浴成为一天之中最享受的时刻。

次卧

全室贴浅米色及金色描细纹的壁纸，搭配贝壳板与珠光白彩纹理的烤漆饰板，做为电视主墙。经典的英式家俱配搭顶级丝质寝具，让身心全然放松舒畅，尽做享豪宅雅室的古典情韵。

次卧卫浴

以湛蓝透白的百合花做为此空间的主题，简单的磁砖色彩配置更放大了卫浴空间，进口精品的按摩浴缸正等待主人卸下心防，沉浸其间。满贴的大面镜子及精致的画框，加上雷射切割板饰，更提升卫浴整体的精致感。

孩卧

电视墙上白底浅蓝印花壁纸，不同于主、次卧较沉稳风格的湛蓝几何线条，孩卧的设计较为活泼且带着天真童稚的情趣。利用床背的镜子做简单的造型，在大面镜子分隔线条中再嵌入较小的方形彩镜，让古典造型的床座背后，展现较成熟美丽的背景。梳妆台侧的收纳柜面、柜身皆以白色烤漆处理，减轻视觉负担。

公用卫浴

全室皆以手工拼贴马赛克壁面，白底与纤细的金色与银色交叉弧线形成雅致的拼花。一贯采用的顶级精品卫浴设备自不在话下，下半墙面搭配使用浅金锋石材，气势恢弘，更让整体客用卫浴风格提升为总统套房的等级。

Household with buildings of four kinds of room the Daping number, pastel Victorian interpretation of visual effects enlarge the space, but also show this peerless luxury, refined and elegant.

Shanghai in recent years emerged as the world-class city, with the central urban area of land available for development dwindling, the demand trend is more obvious, excellent planning of the mansion is a house is hard to find. Far in Fenghua location is very superior, is located in Jing'an District, the core area from Nanjing West Road business district just two blocks. Quiet, lots of rare, planning the whole, together with the Far Glory Group extraordinary construction quality is superior, other residential in this area is much higher can not hold a candle to. As the unparalleled luxury of such disdain for the group owner, the May 2006 opening of the far-stroke Hua Yuan has launched four of a total of 292 suites source, from the price point of view, this year, reported 35,000 yuan / square meter are The price has set a benchmark price of Jing'an District. Desire for the taste of life, higher in the pursuit of the ordinary mortal.

Two important elements of lifestyle, value appreciation, the other is the style of living practice of power. The basaltic design Victorian-style design in this account is, of course, because of its decorative elements in the arts, far-reaching impact; but also because this style to enhance the aesthetics and taste, it is precisely in line with the new Shanghai vigorous take-off which gave birth to a new value. Its use of color is bold and brilliant, strong contrast, neutral color with brown, gold highlights the luxury and amplifier; its shape and delicate, sophisticated space partitioning, rich layers of decorative beauty of the natural beauty perfect combination, it is true aestheticism reflected. As a result, the Victorian style is still the source of many of the design and creative elements, is an elegant model of five-star hotels and the magnificent manor houses often used.

Far in elegance the SH3-8 large basaltic design use ingenuity to capture the beauty of the classic Wedgewood Royal Ceramic Art design tone, fully reflected in the Victorian art of living.

Spatial detail performance, abandoning the magnificent over-decoration, shaping a leisurely, delicate aristocratic taste. White substrate, rendering the soft colors: light blue, green and beige, like Wedgewood's famous jade relief carved and detailed, more emphasis on the extravagance and three-dimensional sense of space.

Join the line of plates, hollow, window grilles and other decorative elements on the ceiling and wall, graceful lines to create the space atmosphere of elegance, the fabric soft interface, such as: drapes, wrap blanket, pillow, bed decorated with silk and velvet, etc. The shiny fabric to create a royal style. Patterned floor mosaic and carved window lattices, from the simple and honest material to show the artistic temperament, but also a better microcosm of the era of retro spirit.

Flashy light and shadow of the beautiful new realm - the public space of the collection of royal style

Set the tone of public space-based households aristocratic and elegant style, including the entrance, dining room, kitchen. The overall use of color is light beige and white, with pale gold, silver, and a few black strokes, and elegant light sky blue. Relief opened the entrance door, just as the stepped into the beautiful new realm in the Wardrobe, isolated from the outside the hubbub of the world, suddenly enter the castle palace of the royal style. Medallion presented by the entrance floor of beautiful letters ornamentation, as the general emblem of the royal family marked the masters of noble status, and also notice a full account of Victorian decorative elements and elegant style.

The main part of the public space begin with elaborate floor stone

decoration focuses on color, decoration, lines, texture, and gloss through with each other against the background of these decorative elements to create a fine china fine carved texture, showing Shanghai upstart The mansion style degrees.

Entrance

Basaltic use of design ingenuity, in in Farglory the construction FARGLORY English name, and retrieve the F and G of the English letter combinations, stone mosaic ground floor to create the momentum of the royal family crest, but also represents the joy of the construction industry for housing master must Welcome to Chen.

Greets, modern and traditional craft techniques (laser cutting totem Stickers silver piano lacquer box decorated with the combination of sheet metal and glass) together to create an elegant screen. Welcome the special shoe for the right hand side door piece gray mirror combination of elegant colors and custom door frame, a testament to the low-key and luxurious design intentions.

Parlor

Nearly six meters long main wall of the large-scale TV, Italian white travertine as a visual subject, the natural pore lines, emphasizing the profound and solid sense of history. The middle and on both sides of the wall by the gold mosaic of mirrors and mirror combination, with a thin silver border, so that the whole piece of the wall is gorgeous but not gaudy.

Sofa back wall of the basement of a warm gray wall covering with different paint thickness strip, such as frame-like cream-colored wall covering mounting of them. In the light irradiation, has further deepened the sense of three-dimensional of the mural.

The specially selected combination of strip of black and gold paint ceiling, piping like to strengthen the visual effects; Can Ya caused the Swarovski crystal chandeliers, custom-made well-known international boutique sofa, texture and luster deep rich celebrity favorite, define gorgeous romantic atmosphere of the living room.

Restaurant

Stone on the floor depicting the classic pattern of fine traditional porcelain Wedgewood, two meters in diameter of the circular mosaic totem directly point out the style theme of the space. Circumference plastic front of the main wall of white travertine a modeling fireplace, below the candle lamps, decorative embellishment. Fine china fine tableware, tea, snack tray, and even the elegant tea table decorations from the table, bring out the Victorian art of living, the mistress of the elegant figure seems to be able to see the shuttle during leisure time in an English afternoon tea Yan to carry out.

The main wall flanked by two pairs of door, elegant lines that delineate the space with a round silver hollow board, clever public space separated from the privacy of the rear.

Kitchen

Extension is the same token totem of the mural, as well as the large area of clear glass doors from the restaurant into the kitchen during the day, the balcony light into.

The kitchen is equipped with the top custom quality kitchen island, open planning, cooking and as every day most enjoyable experience contented.

Elegant if the dream of the soul Imperial Garden - elegant and restrained private space

Walked down the aisle, left, right after each entered the range of private space. In this space, the same strain of course, is the Victorian Royal style, but more elegant and delicate aesthetic attention to personal taste sway.

Private space, extensive use of different lines and texture of the wallpaper, mosaics, mirror and strip, tiles, stone inhabit every family, to build the stage of their individual tastes. Imperial Garden where the mansion house in the heart, people forget the hard work day in a beautiful pattern of retro pure.

World-renowned designers Frank de Biasi says: "Design is the combination of creativity and inspiration, is to develop the potential of a room or a home." The basaltic design in this space in the house give full play to the creativity and systems work together, so that every celebrity owners to see the unique beauty of their dreams of belonging to the family home, originally through the inspiration and ingenuity of design, ahead of schedule.

Aisle

The walkway between the public and private space at the junction of moderate to guide the turning point of the road moving lines and mood.Two sliding doors hidden in the restaurant after the whole piece of the aisle labeled champagne gold color interlaced box wall, walkway lights lit when the restaurant banquet guests to see behind the shining faintly through the carved doors murals, woven into a beautiful window view of a beautiful symmetrical indoor.

Den / Living room

Use the module to unit murals combination of wall, den / living room also has a side stand wall shelves, study as the traditional British style. Different colors and materials, designers make extensive use of a white sheet and floor mirror each other against the background of the mirror and book wall, seems to be back with love and beauty of art, to reconcile the material and spiritual Victorian literary era.

Sound of the piano like a dream, scholarly light diffuse, whether it is reading literature Jie Qiaozhi Eliot (George Elliot) in this novel, or taste the love poems of Robert Browning, people could not help germination to ponder times past.

The master bedroom

Blue pattern wallpaper with a high waist white strip around the entire space, to create a comfortable atmosphere for the bedroom due. The main wall of the bed back to eye-catching and dazzling peacock feather collage wallpaper, indirect lights and shiny texture reflecting the varied and splendid.Classical amplifier King Size bed Tower British import collection of furniture, elegant soft lines of fierce determination, a symbol of the status of the host and hostess noble and extraordinary character of taste.

Master bedroom dressing room

Extension of the blue hue to the locker room, cabinet linear skeleton rather than the board assembled Large Stickers-length mirror, to improve penetration, the more magnified the depth and breadth of the space. Armani, Chanel, Hermes and other clothing and accessories can be host and hostess,

after their kind, have star-like custody and care.

Master bedroom bathroom

The whole room wall mosaic handmade collage to show the white gradient totem and the Chinese blue and white porcelain blue pattern, mix and match the Chinese compromise classical aesthetics in Western elegance. Building materials with a little gray lines of the stone is white, gold mirror of the entire film mirror match, and champagne silver foil strip decoration box to create a classical mirror mirror, the same boutique hotel is the top imported sanitary equipment, so that bathing to become the most enjoyable moment in the day.

Second bedroom

The whole room, the wallpaper of light beige and gold depiction of fine lines, with the shell plate and the pearl white color, texture paint trim the main wall as the TV. The classic English-style furniture to match the top silk bedding, body and soul totally relaxed comfortable, and enjoy the classic luxury elegant room Charm.

Second bedroom bathroom

To this end space theme to blue through white lilies, simple tile color scheme to enlarge the bathroom space, and imported boutique jacuzzi are waiting for the owner to remove the psychological barriers are immersed meantime. Omo full affixed mirror and exquisite frame, coupled with laser cutting board decoration, but also enhance the exquisite sense of the bathroom as a whole.

Child lying

TV on the wall white light blue printing wallpaper, different from the main, second bedroom a more sedate style blue geometric lines, the child lying on the design is more lively and with a naive childish delight. Bed back of the mirror to do a simple shape, and then embedded in a small square color mirror in the the Omo mirror separated lines, the classical shape of the bed behind to show a more mature and beautiful background. Admission counter, dresser side of the cabinet body to begin with white paint handling, and reduce the visual burden.

Public bathroom

The whole room to begin with handmade mosaic wall tiles, white with delicate gold and silver cross arc to form elegant parquet. Consistent top quality bathroom equipment is self-evident, the second half of the wall with the use of shallow KIMPHONG stone, momentum, leaving the overall guest bathroom styles enhance the level of enjoyment for the presidential suite.

田园美境，法式乡村
The Pastoral of French Countryside

设计公司：达特思室内设计（北京）有限公司
设计师：翁伟锴
面积：360 m²
主要建材：天然大理石、仿古瓷砖、混油木饰面、拼花米地板

Design Company: Details Design Consultants
Designer: Weng Weikai
Area: 360 m²
Materials: Natural Marble, Antique Tile, Wood Veneer, Parquet

    对本样板间业主的要求是要结合建筑的整体风格以及突出度假休闲为旨的特点，在反复的探讨过后，本案最终以法式乡村风格为主题。

    该样板间在设计上还结合了法国南部与西班牙室内色彩的鲜明材质，用明快的色彩营造空间的流畅感，曲线的运用使整体感觉非常优雅、尊贵而内敛。蓝天白云，鸟语花香，普罗旺斯的乡村之美，在法式乡村中完美呈现，配色上的大胆鲜艳便是一大特征。本案中，浅绿色、黄色、奶白色、蓝色等活泼的色彩搭配，反映了乡村丰沃、富足的大地景象。草绿色的墙壁从客厅餐厅一直往上延伸到二、三楼的走道楼梯空间，家庭厅使用了绿底色的碎花和条纹壁纸，而女孩房则毫不犹豫地缀满了柔美的花卉图案。

    而乡村天然粗犷的特点，则用来强调空间顺序的完整性和连续性。通过门廊、门厅、客厅、餐厅、庭院等一系列空间的有机组织和合理安排，表现出一种休闲雅致的礼仪感和价值感。以各个空间的相对独立与秩序感为特色，各主要功能空间规划合理的布局和尺度，避免空间的大而不当。同时通过设计处理，增加空间的通透性，更强化了室内的整体功能格局。

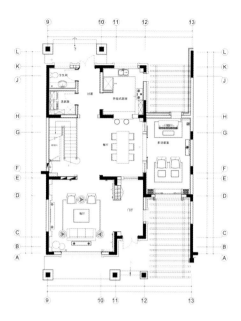

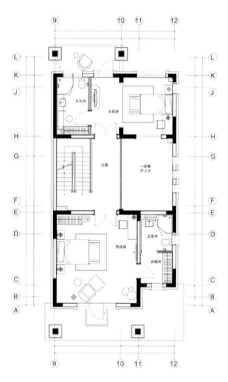

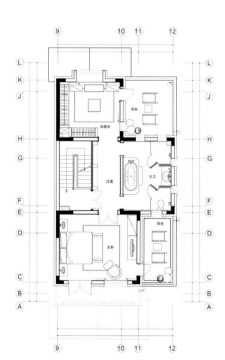

The sample house is aimed to combine the building's overall style to set off characteristics of vacation and leisure, and the adoption of French country style is the result of a reduplicative communication.

The material colors popular in the interior design preferred by France South and Spain shape the fluidity of space, while the use of the curve boasts the overall feeling of elegance, nobleness and restraint. Out of the space, comes an image of a Provence village where the sky is blue, the clouds are white, the birds are singing and the flowers smell fragrant. Patches of light green, yellow, milky white, blue, etc. Reflect a land that's rural, fertile, and rich in sources. The grass green spreads on walls from the living and the dining rooms, to the stair aisle that links the 2nd and the 3rd floors. The family room is decorated with green floral and striped wallpaper, and the room for the daughter is shifted into a flower world.

Rural characteristics exclusive to co untry are used to emphasize the integrity and continuity of the sequence space. With an organic and reasonable arrangement of the porch, the foyer, the living room, the courtyard and a series of spaces, a casual and elegant sense of etiquette and value is thus made. While relatively independent, each space is endowed with an appropriate layout and scale to avoid a space large but irrelevant. And at the same time, the increase of spatial permeability strengthens the overall function of the interior pattern.

倾心自然，舒适回归
Passion for Nature, Comfort to Return

设计公司：达特思室内设计（北京）有限公司
设计师：翁伟锴
面积：230 m²
主要建材：天然大理石、木饰面、拼花米地板、艺术马赛克、手绘壁纸

Design Company: Details Design Consultants
Designer: Weng Weikai
Area: 230 m²
Materials: Natural Marble, Wood Finishes, Parquet Floor, Art Mosaic, Hand-Painted Wallpaper

仿古、沉稳、狩猎、混搭，世界上最大殖民帝国——英国，殖民地领地遍及欧非亚美澳五大洲，有"日不落国"之称，英式殖民风格也就变得包罗万象。本样板间的设计正是采用了英式殖民地风格，摒弃了繁琐与奢华，并将不同风格中的优秀元素汇聚融合。

在设计手法上，以舒适机能为导向，强调了"回归自然"而又恰到好处地把雍容华贵之气渗透到每个角落，使整体空间变得通透流畅，突出了别墅本身的自然优势又适当彰显业主的个人品味。仿古家具铺陈殖民地情调，大幅的18世纪风景画被描绘于玄关以及卧室的墙面上，在木家具的衬托之下，仿佛将人带入怀旧的殖民时代。

客厅经过精心布置，新颖别致的壁炉与墙面两侧的凹龛有机结合，天花仿旧木梁与仿古吊灯相趣成章，给人以庄重豪气之感。整个地面运用了天然石材拼贴，与墙面仿旧肌理质感的壁纸形成对应，以及仿古的沙发、欧式的茶几、台灯，共同营造温馨祥和的气氛。餐厅作为一个单独的空间，地面的简洁拼花，与整个主题相呼应，让每个空间都能感受到设计元素无处不在。藤色和稻色的油漆营造了英式温暖的感觉，而繁复古典花纹图样的壁纸则明显美化了书房的主墙。

Antique, calm, hunting, mix and match the world's largest colonial empires - Britain, the colonial territories throughout Asia, the United States and Macao, Europe and Africa on five continents, the sun does not set the country ", said the British colonial style, becomes all-encompassing. The design of this model is the British colonial style, get rid of the cumbersome and luxury, outstanding elements of convergence in the different styles and fusion.

Design methods, comfortable function-oriented, emphasizing the appropriate "return to nature" and the right elegant of the gas penetrated into every corner, so that the overall space to become transparent and smooth, highlighting the natural advantages of the villa itself demonstrate the owners personal taste. Antique furniture, elaborate colonial ambience, large eighteenth-century landscape painting was painted at the entrance and bedroom walls, wood furniture set off, as if into the nostalgic colonial era.

Use of natural stone living room after the whole ground carefully arranged, organic combination of the novel fireplace and walls on both sides of the concave niche, smallpox Distressed wooden beams and antique chandelier interesting chapters, giving a sense of solemn pride; collage, and wall Distressed texture textured wallpaper the formation of the corresponding, antique sofa, European-style coffee table, table lamp, and jointly create a friendly and peaceful atmosphere. The restaurant space as a single, simple mosaic of the ground, with the entire theme echoes, so that each space can feel the design elements everywhere. Rattan color and rice-colored paint to create a British warm feeling, complicated classical pattern pattern wallpaper significantly beautify the main wall of the study.

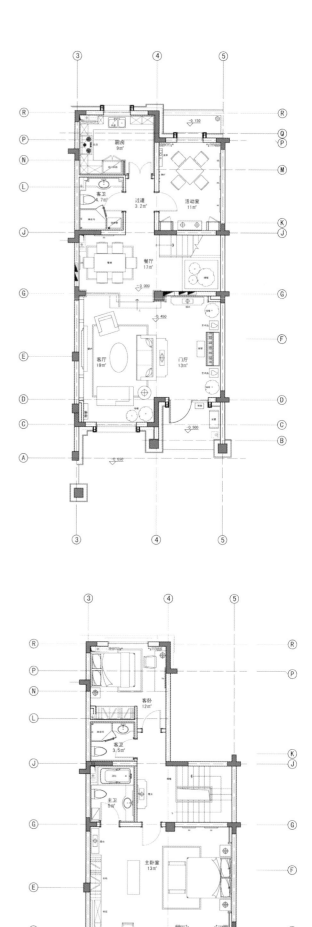

洗净铅华的优雅
Elegance to Be Natural

设计公司：逸品·原宿设计工作室
设计师：王伟

Design Company: Yipin Harajuku Design Studio
Designer: Wang Wei

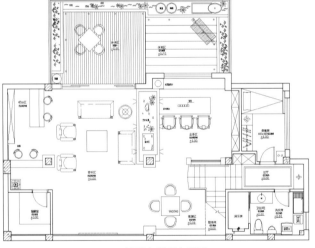

地下室平面布置图

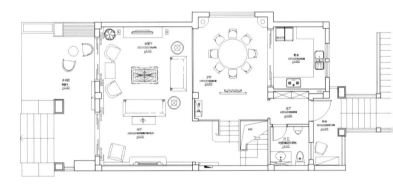

一层平面布置图

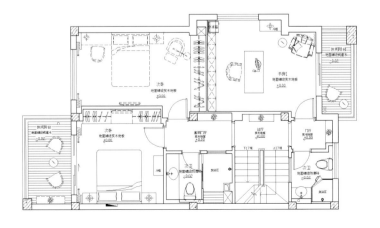

二层平面布置图

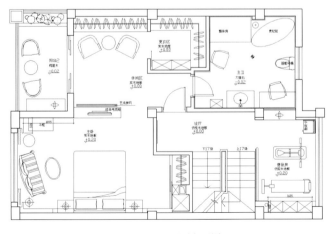

三层平面布置图

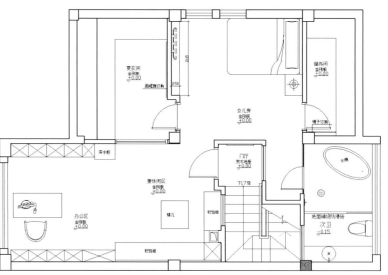

四层平面布置图

  本案在延续古典欧式雅致大气的同时又融入现代简约时尚的元素，力求将空间打造成既富有艺术气息又优雅华贵的新古典居所。在这个空间中设计师充分融入了浓郁的欧陆风情与浪漫的西式情调，并巧妙地糅合了简约主义的爽利、爽洁与传统欧陆风格的精美、绮丽，衬托出高贵优雅的氛围。

  整个大厅通透明亮，雍容考究的仿古砖与墙体线条简练硬朗，给空间以无限的张力。蓝色窗帘的使用更是给人无尽的梦幻之感，从色彩角度诠释了空间的跳跃与渐变，无形中给居家增添了几分神秘与浪漫。

  厚重感十足的真皮沙发端庄大气，与镂空实木案几相映成趣，大面积墙体留白使空间极富视觉表现力，加之室外光线的完美融入，更加凸显出其自然清新、简练优雅的气质。楼梯两侧

的布置颇为中式，精致的博古架上摆放着业主珍藏的瓷器，左侧的案几由一块阔大的实木木板做成，古朴低调，给人极深刻的印象；楼梯右侧是古色古香的中式桌椅，高挑的靠背与稳健工整的椅角形成对比，繁复与精致在此完美结合，与暗红色的格调一起点亮空间的激情。置身其中，让人心生无尽的爱恋与欢愉。

A project this space is a continuation of the classical European style that integrates modern minimalist fashion elements, to provide a luxurious neoclassical residence immersed in an artistic and graceful air, when blending Continental romantic ambience and minimalist charm to bring out the elegant atmosphere.

The lobby is bright and transparent, where wall lines concise and tough, and elegant antique bricks complete an infinite tension, the curtains of blue stimulates a dreamy sense to potentially add to the home mystery and romance when interpreting the spatial jump and gradient change.

The leather sofa, sedate and dignified atmosphere, makes a sharp and intended contrast with the hollowed-out wood table. Large areas in blank allow

for a great visual expression, and once daylight spills, a temperament natural, fresh, concise and elegant is highlighted. The configuration on on both sides of stairs is quite Chinese: the antique shelf is a good collection of porcelain, the tea table on the left side is made of a wide wood plank, simple and low-key but to be imposing, and the tables and chairs continues to be Chinese, whose backrests are high and corners are neat. Contrasting but blended the complicated and the delicate to ignite the spatial passion with the color of maroon. Once here, it's bound to generate endless love and joy.

舒适典雅的贵气
Comfortable, Elegant and Luxurious

设计公司：大勻国际设计中心
设计师：汪晓理、李璐璐
面积：400 m²
主要建材：橡木染色、橡木地板、圣罗兰黑金石材

温情中的恬静

联排别墅，设计着力营造出舒适温馨的美式家居环境，除去繁复的纹样线条及奢华的配饰，在棉麻布艺与温情的橡木中，寻求一种舒适精致而高雅的生活。地面层主体空间延用轴线式对称的空间更显端庄；玄关渐进式的表现，将待客等候空间与正气、高耸的客厅功能区分明确，更显主人的尊贵。12人的西餐桌，结合建筑六角窗放置舒适的下午茶桌、带早餐台的大厨房，都是这个豪宅的机能框架。

不拘一格的气度

二楼、三楼以卧室为主，有大套房与小客房，三楼整层楼均为主卧空间。主卧室以卧室为中心，起居、书房及独立更衣室、卫生间分布在卧室左右两边，奢华大床，以舒适、高级见其生活品质。不拘一格的设计，将阁楼空间定位成女孩房，淡雅的色彩与楼下沉稳的风格稍有不同，年轻、叛逆的气息在空间中回荡，独立化妆台印证女生的蜕变，卧榻式的书桌怀揣着女生千千万万个美妙梦想……

Design Company: Symmetry Design Center
Designer: Wang Xiaoli, Li Lulu
Area: 400 m²
Materials: Colored Oak, Oak Floors, Yves Saint Laurent Black Gold Stone

Tranquility in warmth

The townhouse is destined for an American dwelling warm and comfortable but free of complex patterns, lines and luxurious accessories, where the employment of cotton fabrics and oak seek a life that should be comfortable, refined and elegant. The ground is axle-symmetrical to set off a more dignified nature. And the incremental foyer, when defining areas of the waiting and the living room, highlights the status of the owner more distinguished. The western-style dinner table that can accommodate dinner up to 12 people, together with afternoon tea table by the hexagonal windows and the one for breakfast in the table set off a much clearer functional frame of the mansion.

Eclectic style

Bedrooms on the 2nd and the 3rd floor are respectively large suites

and small guest rooms. The 3rd floor is taken up by master bedroom. A space it is fully facilitated by sections on wings of the master room, like the living space, the study, the independent changing room and the toilet. The bed is of large size and of luxury, whose comfort and high end are bound to reveal a good quality of life. The attic is shifted into room for adult daughter, whose elegant color slightly differs from the calm style downstairs, but overspread the characters young and. The separate dressing table witnesses the growth of a girl into adulthood, while the bed-like desk is a carrier of numerous dreams of an adolescent.

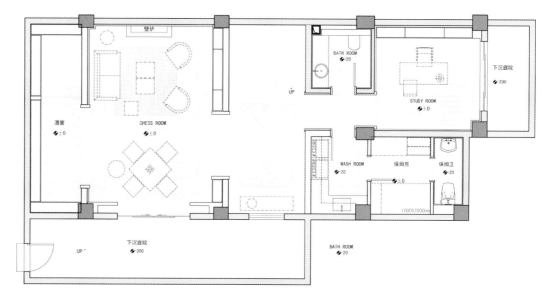

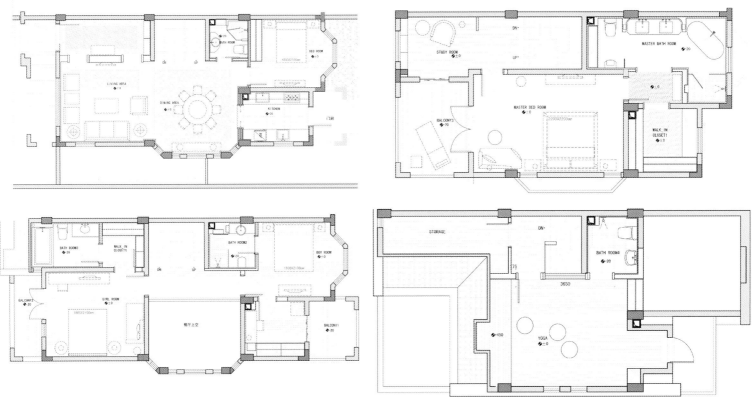

再现英伦精致品位生活
Reappearance of the Exquisite British Life

设计公司：大匀国际设计中心
设计师：陈雯婧、王华
面积：700 m²
主要建材：非洲胡桃、木橡木地板、米色皮革

Design Company: Symmetry Design Center
Designer: Chen Wenjing, Wang Hua
Area: 700 m²
Materials: African Walnut, Oak Flooring, Beige Leather

从挑空大堂到丝绒影音室

　　回归生活的最初点，本案试图用现代的设计手法阐释古典英伦，在原有传统英式住宅空间格局下，以蓝、灰、绿富有艺术的配色处理赋予室内动态的韵律和美感，挑空的大堂及舒适的餐厅配以舒适的大尺度美式家具及手工质感的小饰品，更显品位。以红色为主色调的地下视听室，采用丝绒质地将整个空间烘托的更妖娆多姿。周边设有台球活动区域，让整个空间与主人更好地互动起来。

此处即是秀场

主卧室强调空间的层次与段落,作为主人的私密空间,主要以功能性和实用舒适为主导,软装搭配上用色统一,以温馨柔软的布艺来装点。主卧配套的更衣室,将奢华大气演绎到极致。不在巴黎,也不需前往米兰,此处即是最华贵的秀场。

阁楼塑造的梦想之城

设计师将空间赋予更多的生活化,我们将这个造梦的阁楼,设定成女主人的多种用途的心灵空间。女主人作为服装设计师,在这一亩天地中,尽情发挥自我灵感,在忙碌过后,内设 SPA、美容、YOGA 区域,更可带来放松心灵的无尽体验。

From the lobby high-ceiled to the video room velvet

Let's return to the origin of life; this project is a trial to interpret the classic British style with modern design means, where based on an original and traditional British special layout, the artistic colors of blue, grey and green endow the room with the sense of rhythm and beauty. The high-ceiled lobby and the comfortable dinning room are brought out a more refined taste when equipped with comfort American furniture of large size and hand-made accessories. The underground audio-visual room in a red hue is set off more beautiful by the velvet texture. With the billiards area surrounding, a better interaction between the entire space and the owner is ignited.

Here is a show field

The master bedroom features a sense of layer. As the private space of the owner, it is marked by great functions and comfort, with furnishings unified in hue and fabrics timely to ornament. The changing room built in the master room perfectly interprets the luxurious atmosphere to the utmost, making a most resplendent show field accessible only in Paris or Milan.

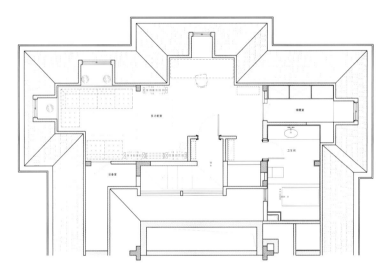

A dream city shaped by attic

When the whole space is life-oriented, the attic is transformed into an ideal place to cook up inspiration, where the hostess, a professional costume designer, can enjoy a spiritual journey through areas of SPA, cosmetics and YOGA.

中西合璧，混搭创新
Innovation by the East and the West

设计公司：LSDCASA
设计师：葛亚曦
面积：153 m²

Design Company: LSDCASA
Designer: Ge Yaxi
Area: 153 m²

"混搭"，已然是设计界最炙手可热的手法之一。但是，混搭并不是简单地"拿来"。在LSDCASA看来，混搭不是乱搭，是懂得选取互容风格的元素，只有对各国的历史和文化有深入的了解，才能使混搭更加出彩、出位。

这套居所的硬装设计出自美国BLD的手笔。提到BLD，中国的样板房设计圈多少有点敬而远之。其频频见诸于酒店会所的设计行列，但是沉稳、内敛的风格，和国内追求华丽、霸气的"豪宅"市场显得格格不入。他们一旦在样板房露面，就会让后期的软装设计觉得无计可施。

整个空间在硬装上全部采用了低调、硬朗的直线条，连墙纸都看不到任何的图案，用手触摸才能感觉出细微的肌理。如何让这样一个低调的空间高调亮相，LSDCASA的解决方案便是：经典意大利混搭泛东方文化，将意大利文艺复兴时期与东方明末清初时期同样是皇室的物品完美地融合，打造出立足于艺术与文化，却立于预料之外的"个性"豪宅。

进门玄关处，费拉门戈歌舞挂图首先带来一份轻松的心情。在主人的起居室，混搭在这里拉开了序幕，看不到任何华丽的装饰，看不到任何具象的堆砌，一切都仿佛是经历了时间的沉淀，从历史的昨天延续到了今天。沙发和茶几一看就是意大利时期的产物，两把兽头扶手椅的中间，安放了一个很中式的案几，呼应壁炉上的花瓶、陶罐，以及墙角的木箱，为空间注入一抹中式的色彩。天花的吊灯特别选用了云石和铜的材质，摒弃了轻浮的味道，让整个空间的气质更显沉稳与大气。

在主卧间，LSDCASA将硬装与软装进行了巧妙的结合。可以看到，墙上的木饰面，在后期的家私上得到了很好的延续。蓝色则是这个空间中一种出彩的点缀，它们出现在不同的角落，让空间显得不是那么的单调和沉闷。床尾这个特别的电视柜是根据清代的家私进行了现代的改良，恰到好处地烘托出了空间的品质感和艺术感。

"Mashups", already one of the hottest design community practices. However, the propriety of a good grasp of mix and match is not simply "used". Seems LSDCASA, mix and match is not chaotic ride, is to know how to select the elements of the style of the mutual capacitance, only a deep understanding of the history and culture of the countries in order to mix and match more color out.

Hard-mounted design of the set of ownership from the United States BLD's handwriting. Referred to the BLD, the Chinese model of the design circle somewhat at arm's length. Frequently seen in the ranks of hotel club, but a calm, restrained style, the pursuit of gorgeous and domestic, vested with the "luxury" market looked out of place. They appeared in the model room will make the post-soft-mounted design feel can not do anything.

The entire space in the hardware installed on all low-key, tough straight lines, even the wallpaper do not see any pattern, to touch to feel the subtle texture. How to make a high-profile debut of such a low-key space, LSDCASA solution is: mix and match classic Italian pan-oriental culture. The Italian Renaissance and Oriental Ming and Qing period, the same items of the royal family is a perfect blend to create a "personality" based on the arts and culture, but stand unexpected mansion.

Door entrance at sliding Ge Fei Song and Dance wall charts, first of all bring a relaxed mood. In the living room of the master mix and match here opened a prelude to see any ornate decoration, to see any concrete piling up, everything seemed to have gone through the time the precipitation from the history of yesterday, continues today. A sofa and coffee table is the product of the Italian period, the middle of the the two Shoutou armchair, placed a Chinese case a few echoes of the vase on the fireplace, pottery, as well as the corner of the wooden box space to inject a touch of Chinese

color. Ceiling chandelier specially selected materials of marble and copper, get rid of the taste of the frivolous temperament is more calm and the atmosphere of the entire space.

In the master bedroom space, LSDCASA will be hard-mounted to a clever combination of soft loading. You can see, a good continuation of the wall of wood finishes, the furniture of the late. Blue is the embellishment of this space is a color, they appear in different corners of the space it is not so monotonous and boring. The end of the bed this special TV cabinet is based on the Qing Dynasty furniture, modern improvements, the right to express a sense of quality and artistic sense of space.

英式下午茶
Afternoon Tea

设计师：周光勇
面积：260 m²
主要建材：意大利蜜蜂瓷砖、法国之家家具、德国 haro 地板、布鲁斯特壁纸

Designer: Zhou Guangyong
Area: 260 m²
Materials: Italian Bees Tile, Furniture of France, Germany Haro Flooring, Brewster Wallpaper

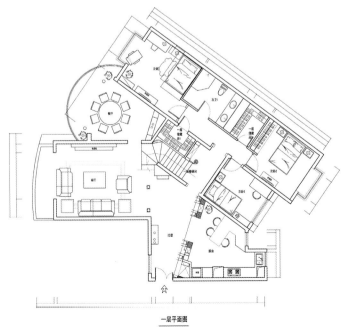

一层平面图

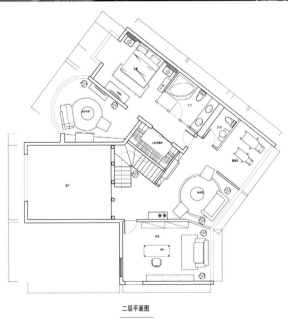

二层平面图

一个家的风格,更多地表达出来的是主人多年来对某一种生活方式的向往。整套设计处处体现新古典风格的特征,罗马柱的磅礴气势、大理石壁炉的贵族情结、圆形穹顶的浪漫辉煌、护墙板的精致细腻……精致而优雅。

在与业主沟通后,选择了卡其色的墙漆与白色的护墙板搭配作为整个公共空间的主色调,稳重而不失轻盈。吊顶、廊柱都选用了新古典典型的式样,但根据层高对其线条搭配进行了简化,既将整个空间勾勒出古典气质,同时又规避了房屋结构和层高的局限。整个空间奢华而恬淡,犹如英式下午茶般精致典雅。

后记

在秋日的一个午后,阳光洒满了屋子的每个角落。游走在房子里面,细细端详优雅精致的白色梁柱与各色物件,感受着各种细碎小花与格子的温和细腻。卡其色的墙,折射着阳光,恍如咖啡面上那层诱人的奶沫!

Out of the style of a dwelling space, comes mostly the desire for a lifestyle in the depth of the occupant's heart. And everywhere in this project, reflect neo-classical features of marble pillars, marble fireplace, dome and wainscot boards.

The communication with the owner finally results the hue of public space that's made by khaki paint and white wainscot board, prudent and also lithe. Neo-classical, the suspended ceiling and the pillars are simplified in lines that's consistent in height, outlining a classical temperament and breaking away the limitation of the original building structure and height. Luxurious but indifferent to fame or gain, the whole space keeps as elegant and graceful as that of afternoon tea.

Postcript

Sunlight spills across on an autumn afternoon. A meander within would acquire an elegant and refined enjoyment and appreciation of white beams and colored objects, as well as the warmth and the delicacy of varying plaids. And the sun refraction by the Khaki wall is tempting and attractive like milk floating in coffee.

和弦悠扬
Chords Melodious

设计公司：福州华悦空间艺术设计机构
设计师：胡建国
面积：300 m²
主要建材：防古砖、大理石、壁纸，玻璃、实木花格、手工银箔

Design Company: Fuzhou Huayue Space Art Design Institutions
Designer: Hu Jianguo
Area: 300 m²
Materials: Antique Tile, Marble, Wallpaper, Glass, Wood Grillwork, Handmade Silver

　　本案为建筑面积300平方米的单层住宅空间，使用空间较大，在充分满足使用功能的前提下，装饰风格定位为后现代新欧式混搭，目的是突破以往惯性思维模式，做一些大胆的尝试，在主线条为简欧造型的思路下，采用带有强烈现代气息的材质来演绎，如不锈钢加皮革的搭配，镜面玻璃加欧式卷纹的呼应，木作镂空隔断与手工银箔的组合，无不体现强烈的现代工业气息，色彩搭配强调简洁明快，线条造型上硬朗而不失柔美，装饰元素既延续了欧式风格特有的柔美和谐，且不失新颖独特的气质，仿若跌宕起伏的音符共同演绎出一曲和谐的乐章。

　　A single-storey dwelling space of 300 square meters this project is. The spatial abundance, when fitting the necessary needs, is positioned a mashup style post-modern and neo-European, deliberately to break through the usual mode of thinking to make a bold try by interpreting with the aid of very modern materials in a setting of simple European style, like the match of stainless steel and leather, mirror with European weavy grain, the combination of wooden partition and handmade silver foil. All reflect a strong modern industrial flavor, particularly when colors are stressed simple and neat, lines are hale and hearty but without losing the soft sense, elements continue the gentle harmony of European style, and yet new and unique temperament is reserved. And it's no wonder that a harmonious movement seems to be going with, notes up and down.

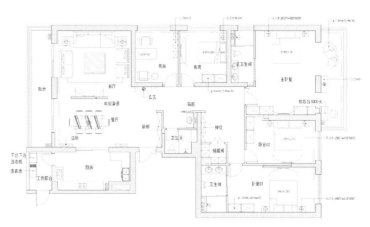

华典悦章
Huadianyuezhang

设计公司：福州华悦空间艺术设计机构
设计师：胡建国
面积：500 m²
主要建材：防古砖、大理石、壁布、玻璃、马赛克、古典油画

Design Company: Fuzhou Huayue Space Art Design Institutions
Designer: Hu Jianguo
Area: 500 m²
Materials: Autique Tile, Marble, Eall Covering, Glass, Mosaics, Classical Oil Painting

　　欧式室内装饰空间最为重要的便是营造出整体的氛围感，并非只是单纯通过刻意的元素追求，而忽略了建筑本身，这要求设计师在空间的构造与细节以及整体空间的驾御上都要进行最为优化的搭配与取舍。

　　这是一个建筑面积为500平方米的独栋别墅空间，复古简欧风格在视觉表现上低调优雅，卓尔不凡，以古典欧式大气的格局感和柔软细腻的线条来营造古典的氛围感，使整体空间在有限的面积里放大延伸，充盈通透。在这个空间中设计师充分融入了浓郁的欧陆风情与浪漫的西式情调，并巧妙地糅合了简约主义的爽利，爽洁与传统欧陆风格的精美，绮丽，衬托出高贵优雅的氛围并有意识地回避了表面的浮奢，涵而不放，只为诠释典雅居住空间新理念。

European interior decoration of the most important space is to create a sense of the whole atmosphere, not simply through a deliberate element of the pursuit, while ignoring the building itself, which requires that should be carried out on the constructed and the details of the designers of space and overall space of controllingthe most optimal mix and trade-offs.

This is a construction area of 500 square meters of single-family villa space, retro simple European style in the visual performance on low-key elegance, Thatcher, a sense of the pattern of the classical European atmosphere and the soft and delicate lines to create a classical sense of

the atmosphere, so that the overall zoom extension of space in a limited area, filling transparent, designers integrate fully into the rich European style and the romantic Western sentiment in this space, and a clever blend of minimalism, physically fit, relaxation and traditional European style beautiful, beautiful, and bring out the elegant atmosphere and consciously avoided the surface of the floating luxury, Han hold only new ideas for the interpretation of elegant living space.

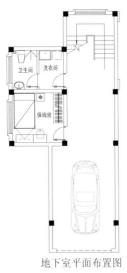

地下室平面布置图

一层平面布置图

二层平面布置图

三层平面布置图

一个空间，两种表情
One Space, Two Expressions

设计公司：活设计国际创意有限公司
设计师：张顺云

Design Company: Live Design International Creative Co., Ltd.
Designer: Zhang Shunyun

古典风格经过历史的锤炼，时至今日，有了另一种时尚的诠释，看似怀旧却又潮流感十足，年轻的屋主对于古典的造型语汇、线条，有着特殊的情愫及难以言喻的迷恋，但表现的方式已跳脱博物馆式的古老复制，融合了个人的喜好及品位，衍生新潮的居家氛围。

在本案中，设计师运用古典的元素混搭现代的手法，或夸张、或对比、或简化，重新定义了刻板的古典印象。空间如同大胆又纯粹的调色盘，设计师引领着屋主及家庭成员完全依照自己的感觉，恣意呈现心中的色彩。

在空间的运用上，除了必要的机能考量，也做了独立的区隔，整个空间是通透且流动的，天花造型及地坪铺面的设计兼具了造型美感，同时也明确划分了空间的属性。一楼的玄关、客厅、餐厅及厨房完全连贯，期望塑造一种独特的大方器度，让来访的客人充分感受到屋主在事业及生活上所展现的企图心。现代奢华的 lounge 氛围、冲突感十足的抽象化巴洛克家具，黑、白、银色系列刻画出的强烈风格，再与屋主的虔诚又外放的双子座性格相呼应。

随着带有幻魅装饰风格的楼梯拾阶而上，眼前又是与楼下完全相反的情景，

纯白的衬底，活脱脱的色彩及家具，不由得对设计师戏剧性的空间操作手法发出惊叹，完全是大胆颠覆的作为。而起居室刻意铺陈的叶片造型地毯，现代混搭呈现的绿草如茵，令人会心莞尔。一体的壁炉墙面，左右对称造型（设计）分别将 minibar 及长亲房入口隐茂整合于其中。

延续二楼的清晰，来到四楼主卧室，一开门，无法拒绝的甜蜜浪漫迎面而来，令人怦然心动，毫不遮掩幸福流泄溢于整个空间，时髦的桃红色搭配订制设计白色的洛可可，像是情人分享着白色礼盒中甜得化不开的蜜糖。

在本案中所有的家具、灯具均是设计师与业（屋）主于国外配色的订制，设计师运用柔软的造型及色彩语汇去传递不同于一般的空间张力，不仅效果显著，也充分满足屋主对家的期待。

The historical quench of the classical style makes it fashion, but still full of nostalgia sense. Though completely free of the ancient copy of the museum style, the classical style involved within is integrated with personal taste and preference for classical shape and lines to derivative a leading home atmosphere.

A project it is to make mashup of classical elements and modern techniques, or exaggerated, or contrasted, or simplified, to subvert the classical stereotype. The palette boldly used contributes to a colorful imagination in accordance with the occupant's feelings.

The layout is partitioned definitely with necessary functions met to allow for an airy and transparent space. Design for the ceiling and flooring are aesthetic and acts as a clear division of the spatial properties. On the 1st floor, the entrance, the living room, the dining room and the kitchen are completely coherent, aimed to shape a unique and generous momentum, so that guests can fully feel the ambition of the owner towards their careers and life. The lounge of modern is fixed with abstract Baroque furniture, where hues of black, white, and silver depicts a strong style while echoing with the Gemini character of the owner.

Stairs winding decorative, charm and magic, leads to a space that makes a complete contrast with downstairs, where the substrate of white purity, colors and furniture, confides in a treatment approach dramatic, stunning and imposing, imposing a completely bold subversion; the leave-patterned carpet in the living room, makes a green and pleasing scene. The fireplace of whole pieces wall, is symmetrical and respectively mounts the doors of minibar and room for the elders on the right and left.

The master bedroom on the 4th floor is a clear continuation of the 2nd floor, where the color of pink is used with white rococo, like two people in love sharing sweet in a white gift box. The entry is bound to strike an immediate sweet and romantic sense.

All the furniture and lamps are bespoke. The soft shaping, the colors and the vocabulary convey a different spatial tension, not only making a significant effect but also living up to the expectations of the homeowners for home.

跨入新式奢华疆界
Luxurious Feeling

设计公司：日向设计有限公司
设计师：李直桓

Design Company: Rixiang Design Co., Ltd.
Designer: Li Zhihuan

于本案，李直桓强调未来概念的设计意象，将光藏在材质当中，意在表现空间之中内敛而细致的优雅。一方面延伸场域景深，让视角藉此穿透整个室内景观；另一方面材质及光影的安排，成为连结空间与人的介质，在奢华大器气韵烘托主体之际，相互构筑出一幅栩栩如生的框景，进而塑造出一场华丽奇幻的摩登飨宴。

科技 华丽 未来空间概念

座落在台北士林地区，四层楼的别墅建筑享有环伺山林、流水的自然景观，李直桓希冀打造一座未来概念的城堡，以华丽、迷炫、灿烂的姿态，重新定调奢华。材质、颜色、光影颠覆传统的介面规划，透过极致的奢华表现，创造一种属于本身与自然冲突的对应关系，促使空间产生特殊肌理，使空间透过与大环境合谐共生，激荡出瑰丽、闪亮、奇幻的作品。

地下室车库是收藏爱车的地方，利用镜面材质交错拼贴，营造视角通透效果，同时挹注时尚前卫概念，宛若打造出品牌车系的SHOW ROOM。

一楼客厅区域区分为大、小客厅，大客厅以两盏水晶吊灯灯饰作为空间主轴，电视墙面透过不同颜色的玻璃马赛克拼贴与原始梁柱利用镜面砖包覆方式，相互辉映，陈述着华丽经典的空间故事，同时为接下来各个区域的雍雅精致揭开序曲。

建构奇幻璀璨的极致奢华

拆除原始建物的楼梯，以新设计的楼梯介面为中心，大、小客厅分别座落两侧，利用高低落差，作为回旋梯的主轴概念，扶手以不锈钢电镀、雷射雕花为主，踏板藉由明镜与五彩雷射的层板交错亮丽的表情，楼梯下方表面以玻璃马赛克贴覆，地材藉由镜面马赛克、LED灯光的光轨设计，规划出上下对应的旨趣。一旁的墙面设计，以钻石切割方式为主的概念墙，汲取适当的自然光束、LED人造光源，交错层叠并调制出未来的科技感受。餐厅成为厨房及小客厅的介面关系，圆形的安排，与天花、餐桌造型呼应，主墙以波浪饰板、钢烤作为收纳机能的介面。

私密空间的表现上，利用墙面作为主题式的串连，让每个空间的隐喻，藉由墙面材质的穿插融合，展现独特专属的风格。二楼壁面以红、金色为主，不同深浅衍生出视觉效果及动线发展。男孩房以黑、银相间的壁纸，营造沉稳、内敛的氛围；长亲房以白色为主，延续至起居空间，企图营造脱俗、优雅的卧眠意涵。主卧区域以灰镜为介质，通透明亮，明镜、大花图腾的壁纸、马赛克反射空间璀璨斑斓的质感，木质地板镶嵌不锈钢饰条，延展时尚意象，与起居室以电动门界定，加深空间细致的感官印象。

305

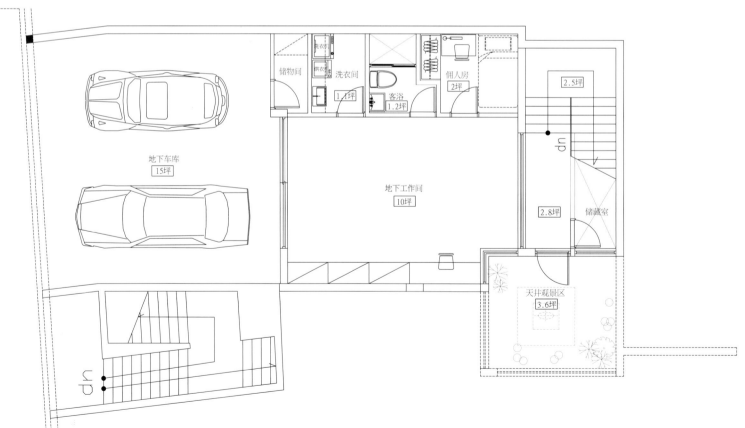

地下室平面布置图

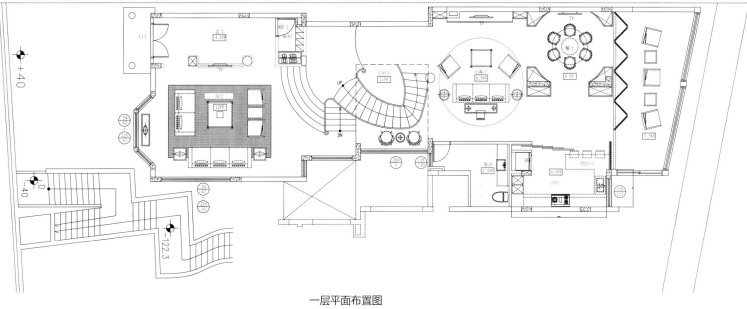

一层平面布置图

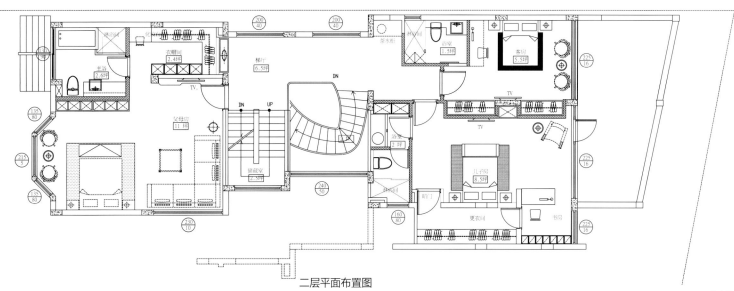

二层平面布置图

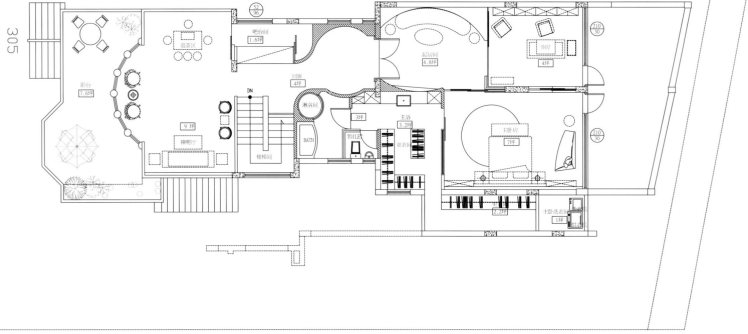

三层平面布置图

A project this space is to stress an image of future concept, to hide light in the material with an aim of expressing the inherently restrained and the meticulously elegance, and to make a field extension to help perspective penetrate the interior. With the aid of arrangement of materials, light and shadow to link the space and the media, the luxurious render an amplifier to express the spatial hero while both are complimented to each other make a vivid frame and shape a scene of brilliant and modern feast.

Spatial concepts in the future: tech and magnificence

The surroundings of forests, mountains, and waters are destined to make a future-oriented castle with its magnificence and shine to redefine luxury. When materials, colors, lights and shadows subvert the traditional plans, an expression of ultimate luxury is involved to create a corresponding relation between the space and the nature, bringing forward a special texture to make a co-existence for space and the environment to create shiny, fantastic works.

The basement garage is of mirror collage, creating an effect of transparent perspective, while stressing leading concepts, as if to make a show room for name brands.

The living room of the 1st floor is divided into two sections, one large and one small. The large one takes two crystal chandeliers as its spindle. And glass mosaic mirror on the TV wall set off with the mirror coat of pillars, telling of luxurious stories and drawing overtures of the areas yet to come into view.

An ultimate luxury

The removal of the original stars makes the newly-designed stars into a central position, flanked with the living rooms large and small to sides. The handrails is of stainless steel plating, and laser carving; the pedals are of mirror and laser

laminates; the space under the stairs and the flooring are coated in glass mosaic; the LED light makes a sharp contrast up and down; the conceptual wall is laser-cut with daylight, LED lights shaping a technological feel beyond the current. The dining room links the kitchen, and the small living room, where the circle arrangement echoes with the module of ceiling and dining room, and the main wall of wavy panels, and stoving varnish also serves as storage.

The private spaces are strung by walls, on each of which the metaphors are interspersed and fused with the aid of wall material, to show a unique style. The walls on the 2nd floor are dominated by hues of red, and gold to derive visual effects and moving lines. The room for the boy is used with black, silver wallpaper, creating a calm, restrained atmosphere; the one for the elders are in white, which is extended into the living space to build up a refined, elegant, sleeping area; and the master bedroom takes as its medium gray mirror, transparent and bright, where mirror, wallpaper of large flower totem, and mosaic all reflect the bright and colored texture, the wood flooring is inlaid with stainless steel trims to extend the fashion imagery, and the electric door serves as definition and deepens the detailed sensory impressions.

图书在版编目（CIP）数据

奢尚新古典 / 黄滢, 马勇 主编. – 武汉：华中科技大学出版社, 2012.10

ISBN 978-7-5609-8426-1

Ⅰ.①奢… Ⅱ.①黄… ②马… Ⅲ.①住宅 – 室内装饰设计 – 图集 Ⅳ.① TU241-64

中国版本图书馆 CIP 数据核字（2012）第 236819 号

奢尚新古典	黄滢 马勇 主编

出版发行：华中科技大学出版社（中国·武汉）

地　　址：武汉市武昌珞喻路 1037 号（邮编：430074）

出 版 人：阮海洪

责任编辑：王莎莎	责任监印：秦英
责任校对：熊纯	装帧设计：许兰操

印　　刷：深圳当纳利印刷有限公司

开　　本：992 mm × 1240 mm　1/16

印　　张：20

字　　数：160 千字

版　　次：2013 年 3 月第 1 版 第 1 次印刷

定　　价：328.00 元（USD 69.99）

投稿热线：（020）36218949　1275336759@qq.com

本书若有印装质量问题，请向出版社营销中心调换

全国免费服务热线：400-6679-118 竭诚为您服务

版权所有　侵权必究